P   542

British Tractors for World Farming

An Illustrated History

British Tractors
for World Farming

An Illustrated History

Michael Williams

Line Illustrations by Paul S. Kimpton

BLANDFORD PRESS
POOLE   DORSET

First published in the U.K. 1980

Copyright © 1980 Blandford Press Ltd
Link House, West Street,
Poole, Dorset BH15 1LL

British Library Cataloguing in Publication Data

Williams, Michael, *b. 1935* (*Nov.*)
  British tractors for world farming.
  1. Tractors – Great Britain – History
  I. Title
  629.22'5   TL233

ISBN 0 7137 0959 6

Set in 11/13 pt Monophoto Ehrhardt
Printed in Great Britain by
Fletcher & Son Ltd, Norwich
and bound by
Richard Clay (The Chaucer Press) Ltd
Bungay, Suffolk

Contents

To Judith

Acknowledgements

My thanks to all those who helped in the production of this book by way of information and photographs.

Photographs of tractors were provided by the manufacturers concerned and by the following:

Museum of English Rural Life, pages 8, 28, 29, 35, 36, 39, 59, 85
National Institute of Agricultural Engineering, pages 52, 67, 68, 71, 72, 74, 76, 77, 78, 79, 80, 81, 82 (upper), 83, 87, 88, 89 (lower), 98, 101 (upper)
Peter Shelford, pages 13, 14
Western Development Museum, Canada, page 23
National Museum of Antiquities of Scotland, page 34
Massey-Ferguson, pages 55, 58, 62, 70
Howard Rotavator Co, 32, 40, 82 (lower)
Peter Adams, page 45 (lower), 46
Gerald Lambert FRPS, page 45 (upper)
J. L. Thomas, 'The Sentinel Register', page 50
Bomford and Evershed, page 108
Jasper Spencer-Smith, front cover

Power Farming

Power farming began in Britain when steam engines were first used to provide a tireless alternative to muscle power. For centuries food production had relied upon the physical strength of men and animals, with work rates governed by the pace of yoked oxen or by the endurance of a man with a sickle. In medieval times much of the population had laboured in the fields in order to produce sufficient food for survival. Wind and water power had helped a little – mainly with food processing – but it was steam which offered the first real breakthrough to mechanized farming.

Steam engines were first used on the farm, for threshing, just before 1800, and for the next hundred years British manufacturers led the world in adapting steam engines for an increasing range of jobs in agriculture. The leadership was used to good effect to build up a successful export business, and British engines were usually the first on the farm in many parts of the world.

Nineteenth-century steam engines were exported for many purposes, including powering cane crushers in the West Indies, cotton gins in the United States and threshing drums in Argentina and Russia; cable ploughing in Germany and Hungary; hauling wheat and wool to the railhead in Australia, and operating drainage pumps in Holland and sawmills in France. Many of the top prizes at leading agricultural shows throughout the world were won by British engines and equipment, and the gold and silver medals were displayed with pride at headquarters in Bedford and Ipswich, Leeds and Lincoln, and the many other centres of steam engine production in Britain.

The commercial and technical development of steam power for farming is an extraordinary success story. Companies manufacturing engines and equipment earned a reputation for reliability and quality, and some manufacturers expanded from small, local beginnings to become famous internationally in the farm equipment industry.

This is the background to the start of the farm tractor industry in Britain in about 1896, almost exactly one hundred years after the first recorded use of steam power on a British farm. British tractors quickly established a leading position, both in technical development and in building up an export business.

The first successful attempts to use an internal-combustion engine for farm work were made in America. The Charter Gas Engine Company of Chicago made a prototype machine in 1889, leading to five or six more the following year. These, and others produced by companies such as Case, Hockett, Otto and the Waterloo Gasoline Traction Engine Company, were all based closely on the design of the traction engines they were intended to replace. All the early American tractors were big and heavy, and many of them consisted simply of an internal-combustion engine mounted on the wheels of a steam traction engine. They were used mainly for stationary work, and were never intended for use in the field, where the horse remained unchallenged. It is arguable whether these were really the first tractors or among the last traction engines.

Britain's first commercially successful tractor was of the same type. Commercial success is a relative term, as only one is known to have been sold, but the tractor performed well in public demonstrations, earned favourable comment in the farming press of the day and was awarded one of the highly prized silver medals from the Royal Agricultural Society of England.

This machine was the Hornsby-Akroyd Patent Safety Oil Traction Engine, manufactured in 1896 by Richard Hornsby and Sons of Grantham, Lincolnshire.

The Hornsby-Akroyd Patent Safety Oil Traction Engine of 1896.

The makers announced that there would be a choice of four engines, of 18, 20, 25 or 32 hp, but it is doubtful whether all four versions were built. The engines were made by Hornsby under licence from Stuart and Binney. They were started by means of a blow-lamp and ran on paraffin. The tractor was designed primarily for stationary work, but the gearbox with three forward ratios and a reverse was obviously intended for haulage work.

In a demonstration in 1897 the tractor showed its paces before the judges appointed by the RASE to decide the silver medal awards for the year. The Hornsby proved to be manoeuvrable, and performed satisfactorily in a series of tests, which included crossing obstructions such as railway sleepers and being driven over soft wet soil. In their report on the tractor the judges were obviously comparing the Hornsby with a steam traction engine: one of the advantages they listed for the tractor was 'non-liability to explode'. Another sales point picked out by the author of an 1896 report in the journal *Implement and Machinery Review* was that the engine exhaust had been 'rendered silent' to avoid frightening horses on the road. The *I & MR* report included comments on the labour-saving potential of the new tractor:

'The driver has a good deal easier time of it than in the case of a steam-engine, for there is no fire to be frequently stoked, nor are there any water or steam gauges to be kept under supervision. Indeed, the duties are so comparatively light that one man can easily undertake the driving without any assistance, which, of course, means considerable saving to the user.'

The Hornsby was listed at £500 for the 18 hp version. The price included a waterproof cover, a pair of carriage lamps for road work, spanners, a hammer and chisels, a rear-mounted cable winch and a bucket. The bucket was to supply water for the cooling system, which used 40 gallons a day.

The Hornsby was basically a traction engine, but the machines which followed were tractors, even though the word had not at that time come into general use. They employed the power-to-weight ratio of the petrol engine to good advantage to allow a compact design, light enough for use on the land but with sufficient power to out-perform a team of horses. The men who developed these first tractors showed quite extraordinary imagination, breaking completely with traditional thinking to produce designs which were far in advance of competitors in America or Europe.

Among the first of the new tractors, or agricultural motors as they were usually called, were designs by Professor John Scott of Duddingstone, Midlothian. Scott had been Professor of Agriculture at the Royal Agricultural College, Cirencester. He began work on his ideas for a mechanization system for arable farming during the late 1890s, and early in 1900 he announced the formation of a syndicate to

Professor Scott's tractor of 1903, complete with powered cultivator and seedbox.

finance his development work.

His first tractor was displayed to the public at the 1900 Royal Show near Cardiff. The catalogue entry described it as a 'Motor Cultivator, 20 bhp machine, £375 and upwards according to pattern; smaller sizes proportionately less'. This description did rather less than justice to a design which offered remarkable versatility and was many years ahead of its time. Scott had designed a tractor which included a load-carrying platform at the rear with up to three tons capacity, and which could be used for road transport work at up to 4½ mph, had a pulley for stationary work and incorporated a powered cultivator at the rear with a chain drive, covering a width of 5 ft 3 in. on the principle of the modern 'Roterra'.

The 1900 model was the prototype for further development, and in 1903, when the Royal Show was at Park Royal near London, a new version was displayed. This no longer doubled as a load-carrier, but the powered cultivator at the rear was now combined with a seeding mechanism so that seedbed preparation and drilling could be accomplished in a single labour-saving operation.

The Scott tractors offered features which were quite revolutionary, but an ominous sign of the times was the reaction of the majority of farmers. Professor Scott was already admitting in 1903 that many farmers scorned his ideas and his tractors. There was much more interest abroad, however, and Scott was already negotiating with prospective buyers in South Africa.

Development continued, in spite of the lack of customers in Britain. The 1904 Scott tractor weighed only 26 cwt, which should have been an important sales feature at a time when many arable farmers were greatly concerned by risks of soil damage caused by compaction. The new tractor was a tricycle design with a single rear wheel driven by a Renold-type chain. A special feature of the two front wheels was a device to raise the height of one of them when ploughing. This meant that the tractor could be kept level, even when running with one wheel in the furrow bottom. The tractor had a small load-carrying platform at the rear, suitable for two or three sacks of feed.

The most remarkable feature of the 1904 Scott tractor was a front power take-off. This was gear-driven from the tractor engine and was intended for use with pushed equipment including a binder or mower. The engine used by Scott was a four-cylinder petrol unit, with 3·46-in. diameter cylinders and 4·33-in. stroke. It was rated at 14 bhp at 900 rpm, and the price quoted was only £200.

In 1906 John Scott was invited to speak at a meeting of the Glasgow and West of Scotland Agricultural Discussion Club. He spoke 'somewhat exhaustively', according to the report in the *Implement and Machinery Review*, about the changes which tractors could bring to farming. His audience was apparently composed mainly of sceptical farmers who remained unconvinced by his enthusiastic arguments. *Scottish Farmer* quoted him as follows:

'The agricultural motor is no longer an experimental machine; it has now reached that stage when it can be depended on to do all the work of a farm under everyday conditions, and on all classes of soils, at about half the cost, and in half the time, and with less than one-third of the number of men required to do the same work with horses.'

Scott knew that much of the suspicion and hostility to tractors stemmed from fears of a loss of income for those who bred working horses, and for those who produced the hay and oats to feed them. He tried to soothe the fears of his audience by forecasting, with some accuracy, that the arrival of the motor would not mean the disappearance of the horse. Even when every family had a car, he predicted, there would still be an increased demand for horses for riding.

'As to the idea of excluding the motor in order that horses may continue to keep up the price of corn, hay and straw, it is about on a par with one of the original rules of the Selkirk Farmers' Club, which imposed a fine of half-a-crown on members who did not drink whisky, in order to keep up the price of barley.'

There were some shrewd predictions on the effect increasing numbers of tractors might have on the farm labour force. 'The motor itself will be a great educator and a great maker of engineers on the farm', he said. 'Indeed, the time seems fast approaching when motor power will be universally used on the farm, and farm labourers will know more about motors than about horses.'

Question time at the meeting gave members of the audience their opportunity to criticize and doubt. There were objections to the smell of petrol and engines, and to one farmer the electric motor seemed preferable in this respect. There were doubts about the ability of a tractor to work on hills or on land with stones, and another farmer complained that when a tractor worked in a field it made too much dust. Another questioner was worried by the thought that tractors might help to raise output from farms – and how could they sell the extra? The nearest to a favourable reaction was a comment that tractors might be useful for hauling potatoes to the railway station, sparing horses for other, presumably more valuable, work.

The Scott display at the 1905 Royal Highland Agricultural Show was his most ambitious. He took three different tractors to the show, one of them, according to contemporary reports, powered by steam. At a time when the traditional type of steam traction engine still completely dominated the power farming scene, John Scott outdated it with a paraffin-burning tractor equipped with a tubular boiler. By using paraffin Scott eliminated from his design the bulk of coal and the need for a stoker. His tubular boiler did away with the massive weight of a traditional boiler, and the Scott steamer could be at working pressure from a cold start in three minutes.

Sadly, none of Scott's tractors has survived. They were never a commercial success, and his ideas were unacceptably ahead of their time. Twenty years after he showed the world how a modernized steam tractor could work, American manufacturers were experimenting with tubular boilers and paraffin burners. More than seventy years after the Scott tractors were built the front power take-off was welcomed as a novelty, and some manufacturers were introducing combination machines to till and drill in a single operation.

Dan Albone was a contemporary of Scott, and in some ways he was an even more remarkable and ingenious designer. He was also an accomplished salesman with a flair for publicity and a talent for finding unexpected marketing opportunities.

Albone was the son of a market gardener in the Biggleswade area of Bedfordshire, and there were family connections with the Ongley Arms Hotel in Biggleswade. Dan's interest, however, was engineering, and in 1880, when he was twenty years old, he had set up his own small business in the town making and repairing bicycles. The cycle business prospered, and was stimulated by Dan Albone's personal successes as a racing cyclist. The cycles were sold under the Ivel brand name, chosen because the River Ivel runs through the town. As the business developed Albone was able to build up a considerable export business to the United States, and even to China and Japan. He also diversified, winning a contract to supply axle bearings for the Great Northern Railway.

The internal-combustion engine offered great possibilities to Dan Albone. He built his first motor car in 1898, using a 3-hp Benz engine and a tubular chassis of his own design. The car created a sensation in Bedfordshire, and some local dig-

Dan Albone.

An Ivel tractor ploughing with a binder at a public demonstration.

nitaries received their first taste of the pleasures of motoring as passengers in Dan's car. He built at least one other car, and was also one of the pioneers of motor-cycle design.

Dan Albone's ideas for building a tractor apparently started in about 1896, perhaps inspired by the Hornsby-Akroyd success. He began building his first tractor in November 1901, and the following year he entered it for a Royal Agricultural Society silver medal. The RASE judges decided not to include the Ivel in their awards list because they felt some modifications were necessary, but their report included favourable comment on the design, and praise for Dan Albone who, they said, 'has shown an intimate knowledge of what is wanted in an agricultural tractor, and has produced a machine adapted to rough handling and management by un-skilled hands.'

The original Ivel tractor included most of the design features which were retained throughout the production life of about thirteen years. There were three wheels, and the twin-cylinder engine was mid-mounted, driving through a cone clutch and a single ratio forward and reverse. A prominent feature of the tractor was a large rectangular water tank beside the driver at the rear. This held cooling water for the engine and provided extra weight over the rear wheels.

In its earliest form the Ivel was fitted with an 8-hp engine, but later versions were more powerful, using Aster and also Payne and Bates engines. In 1903, reports in *Implement and Machinery Review* described the Ivel as 10 bhp in the spring, 12 bhp by the late summer, and 14 bhp in the Smithfield Show report of the same year. One of the last production versions, in 1913, was described as 24 bhp.

A new company, called Ivel Agricultural Motors Ltd, was formed in 1903 to provide the backing for production and marketing. Although the company had some distinguished directors and an office in London it remained essentially a small organization, very much under the personal control of Dan Albone in its early years. In spite of its small size the pace of commercial development achieved by the company was almost breathtaking, a tribute to the energy and imagination of Albone himself.

British farmers accepted the new tractor with great caution, and Dan Albone had to work hard for the meagre sales achieved. The export story was much more encouraging, and by 1906 the company claimed to have sold tractors to the Transvaal, Cape Colony, Orange River Colony, New South Wales, Queensland, Victoria, Tasmania, New Zealand, Egypt, Cuba, Algeria, India, the Philippines, the Sandwich Islands, Nigeria, Turkey, Hungary, Austria, Holland, Belgium, France, Germany and Russia – according to the list publicized in 1906. The export successes included a single batch of fourteen tractors for the Philippines, a licensing arrangement with an American tractor company, and a tribute from the 1904 Paris Show, where the Ivel was declared 'undoubtedly the great attraction'. A Canadian farmer was quoted as describing the tractor as 'the new farmer's friend'.

Later, in a demonstration arranged in Italy by the Ministry of Agriculture, observers were so impressed by the performance of the Ivel that a special medal was minted and presented to the company. At home demonstrations were an important opportunity to prove to sceptical farmers that the Ivel could speed up their field work. 'Warwickshire farmers stared in astonishment at an Ivel motor yesterday at Alcester, as it ploughed a double furrow through hard, frost-bound land, in which eight horses could only plough one furrow', reported the *Daily Mail* in January 1904. In 1903 Albone reached an agreement with a Biggleswade farmer to operate what amounted to a demonstration farm. Demonstrations were held on alternate Wednesdays, with an Ivel tractor performing appropriate seasonal work under the scrutiny of visitors. A stock of suitable machines, including a Deering mower, Howard plough and cultivator and an Innes chaff cutter, was held at the farm for the demonstrations.

Some of the demonstrations were arranged as official trials for the Ivel, properly scrutinized and with the results distributed to the farming press. One such trial was held in 1903, observed by invited members of the Institute of Civil Engineers. Six acres of grass were mown in 3 hours 40 minutes, for a cost of 10s 5d (52p) including 1s 9d for the driver's wages (approx 9p) and 5·5 gallons of petrol costing 1s 4d a gallon (7p).

Dan Albone used great imagination in his search for new markets for his tractors.

15

In 1904 he built a bullet-proof version which he took to Bisley, Surrey for a demonstration before senior army officers. The tests demonstrated the cross-country capability of the tractor and proved its resistance to rifle bullets when these were fired at it. The tractor was designed for use as an army ambulance. It had special doors at the rear which could be opened out to form a bullet-proof shield to protect stretcher-bearers carrying wounded from the battle front. The tractor could also be used to operate medical equipment such as ice makers and water purifiers, and to transport medical supplies during wartime. There is no record of any sales arising from the demonstrations.

Another marketing idea was to use the Ivel as a fire engine, and this was demonstrated in 1905 in Bedfordshire, when the tractor pulled and later powered a water pump complete with a crew of firemen in full uniform.

In 1903 a special 'low profile' version of the Ivel was produced at the request of a Tasmanian fruit grower, who had ordered an Ivel to cut the grass under his orchard trees, and two years later a special 'narrow' version was exported to France for a vineyard owner who wanted to work the tractor between his vines. The standard Ivel was 5 ft 4 in. wide, but this was reduced to 4 ft $3\frac{3}{4}$ in. for the vineyard.

The Ivel tractor, like the leading makes of steam traction engines, accumulated honours and medals as it appeared at shows around the world. In only four years from its first public appearance in 1902 the Ivel collection had reached twenty-four gold and silver medals, including the RASE silver medal, won at the second attempt in 1904.

In 1907 the authoritative Argentine publication *The Review of the River Plate* in an article about tractor development in America and Britain said, 'The most successful agricultural motor yet placed on the market is the "Ivel", made in England, and which has literally swept the board of gold and silver medals for its class at shows since 1902.'

Dan Albone appears, from contemporary reports, to have been a man of great charm and good humour. He was exceedingly popular, and words such as 'genial' and 'smiling' are often used to describe him in show reports, which were usually restrained and factual. Generally even more restrained were the reports produced by the judges appointed by agricultural societies to assess entries for the gold and silver medal awards. An exception was the 1905 report of the judges for the Essex Agricultural Association, who awarded a gold medal to the Ivel against three other entries:

'The little Ivel was a marvel of energy and efficiency reminding one of its inventor, for it was here, there, and everywhere hauling plough, mowers, cultivators, driving a threshing machine, and never out of place, always coming up smiling, apparently as happy and cheerful in its various duties as the inventor himself.'

Dan Albone, with his abilities as engineer and businessman, and with his energy and charm, should be as well remembered in the history of farm mechanization as Ferguson, Fowler or McCormick. But he died, in 1906, before achieving lasting fame.

'Thus passed away, at the age of 46, one of the most popular, cheerful and hopeful men connected with the agricultural engineering trade – one who was noted for his buoyancy of spirit, courtesy and good fellowship', said the report of his death in *Implement and Machinery Review*.

Ivel Agricultural Motors remained in business, but the driving force of Dan Albone was lost. The pace of technical development became so slow that rivals were soon overtaking the Ivel, and by 1913 this was becoming obvious. In that year the Ivel, which had once been almost unbeatable, came last in the important North Kent Ploughing Match trials, and was criticized for poor design by the judges in the Port Elizabeth trials in South Africa. The company slipped quietly out of business in the early years of World War I, after a brief attempt to act as the importing agent for a small motor hoe.

The tractor which won the North Kent Ploughing Match trials in 1913, and which also beat the Ivel and assorted American tractors in South Africa in the same year, was a Saunderson, manufactured at Elstow, Bedfordshire, by a company founded by H. P. Saunderson. He was another of the highly inventive group of pioneers designing tractors in the early years of the century, and he was commercially the most successful.

Saunderson tractors began to attract attention when the first of the 'Universal' series was awarded a silver medal at the Derby Royal Show in 1906. This was the second attempt, as a 1905 entry had been referred back by the judges for detail improvements to be made. Tricycle designs were popular at the time, and the early Saunderson followed the fashion. In other respects it was unconventional, including the use of a 30-bhp engine – much larger than most of its British rivals – and offering drive to all three wheels for better traction.

The design was planned as a complete mechanization system. The tractor could be used as a stationary power unit, or as a general haulage unit with its own rear-mounted removable platform for load carrying. The platform had a claimed two-ton capacity and was designed as a tipping unit. There was also the option of removing the load platform and using the tractor unit as the forecarriage for a series of specially designed semi-mounted implements.

To publicize the versatility of their Universal tractor Saunderson arranged an unusual demonstration in a field at Kempston, Bedford, in 1906. The demonstration started at 2 pm when the tractor began to cut a standing crop of wheat, using a binder. After a short time the binder was unhitched and the tractor pulled a threshing drum into the field. The thresher was unhitched, and the truck body was

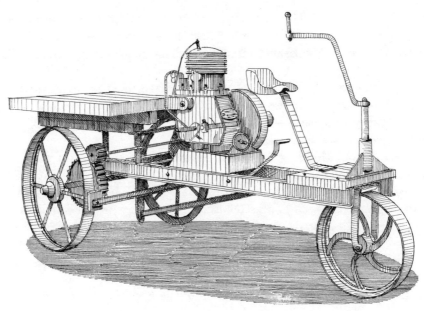

Saunderson three-wheel tractor with load-carrying platform in 1909.

fitted and used to haul the sheaves of wheat to the thresher. Then the tractor pulley was used to drive the threshing drum, and later a grinder to make flour from the newly threshed wheat. Next, the tractor returned to the cleared stubble, and a plough was attached. When the stubble had been ploughed it was cultivated; then a seed drill was hitched to the tractor and a new crop of wheat was sown. Meanwhile the freshly ground flour was made into dough and a batch of loaves was baked. By 7 pm – five hours after the demonstration had begun – wheat had been harvested and turned into bread, and a new crop had been sown.

The 1913 successes were achieved by a later type of Universal tractor, which was completely redesigned. The tractor driver was perched right at the front of the tractor in what is now known as a forward-control design. This was the model G version, rated at 18 to 20 bhp, with a two-cylinder engine and with the facility, added in 1914, to lock the differential to help traction in sticky conditions. In the North Kent trials the winning Saunderson was also the cheapest tractor in the competition, listed at £267, compared to £355 for the Ivel and £1,050 for an imported Stock motor plough, which beat the Ivel into third place.

An addition to the Saunderson range in 1914 was the Little Universal, with a 10-bhp petrol-paraffin engine, and featuring a system for direct attachment of implements with a manual lifting mechanism to raise equipment out of work, operated

Saunderson Model G tractor, 1913.

from the driving seat. The most successful of the Saundersons was the 25-bhp Universal produced in substantial numbers during and immediately after the war. This model, plus an additional smaller version with a 20-bhp engine and a three-year guarantee, were taken over in 1924 by the Crossley company of Manchester when the original Saunderson company went out of business.

James B. Petter & Sons of Yeovil, Somerset were experimenting with tractors before the end of the last century, but their first commercial move into the market did not come until 1902. This was not, strictly speaking, a tractor but a self-propelled portable engine, and it was built as a result of a demand from Petter's agents in South Africa. The engine was a single-cylinder horizontal unit, operating on paraffin, and with $7\frac{1}{4}$-in. diameter cylinder with $9\frac{3}{4}$-in. stroke. It was rated at only 5 bhp, which was developed at 380 rpm. The engine was intended for stationary use but could be used to propel the machine from place to place and pull loads of up to two tons.

A new model, called the Intrepid, was announced in 1903, which was designed for field work as well as for stationary operation. The single-cylinder horizontal engine was described as 12 bhp, with bore and stroke of $16\frac{3}{4}$ in. and 15 in. respectively. Like the Scott and Saunderson designs, the Petter could be equipped with a load-carrying body at the rear.

Petter Intrepid tractor of 1903.

Petter produced an updated version of their self-propelled portable in 1906, and then retired from the tractor market until 1915 when the 'Iron Horse' was launched. This consisted of a tractor unit to be used with horse-drawn equipment. With the Iron Horse attached in place of a cart horse, the driver sat on the implement seat and steered the tractor unit by means of a pair of reins. The start and stop control was extendable to the remote-control driving position. The sales features were the ease of using existing horse-drawn equipment without expensive modification, and presumably also the familiarity of steering by means of reins.

Ransomes, Sims and Jefferies of Ipswich have had a place in the tractor market from time to time, starting in 1903. Their first attempt was powered by a 20-bhp Sims engine with four cylinders, driving through a gearbox with three forward ratios and three reverse. The braking arrangements were designed with unusual thoroughness. There was a foot pedal which simultaneously disengaged the clutch and braked the drive shaft from the gearbox, and there was an additional hand brake to all four road wheels. The three-speed gearing could be applied to the rear pulley, giving a choice of 220, 450 or 1,000 rpm belt speed.

The Ransomes tractor was demonstrated in 1904 at the Suffolk County Show near Bury St Edmunds. The makers claimed it would plough half an acre an hour

The Petter Iron Horse, complete with reins.

with a three-furrow plough, using 1·5 gallons of petrol an hour. They also claimed a drawbar pull capable of hauling an eight-ton load at 7 mph up a 'steep' hill. Ransomes then apparently lost interest in the project, perhaps because of lack of demand in the British market.

Drake and Fletcher of Maidstone, Kent may have received the same discouraging reaction to the tractor they announced and demonstrated in 1903. This was a small tractor, only 8 feet long, with an unconventional three-cylinder petrol engine of 16 bhp. This engine was their own product, equipped with an advanced rotary magneto. The tractor was listed at £350, but was apparently out of production within a year.

Since the demand for tractors was so limited in Britain, Marshall of Gainsborough, Lincolnshire chose to produce their first tractor specifically for export, and particularly to compete with the big American tractors. The 30-bhp model was announced in 1907, and was publicized by means of a 24-hour ploughing marathon in

The Ransomes prototype tractor of 1903.

The Drake and Fletcher tractor from Kent in 1903.

A Marshall tractor of 1909 preserved in a Canadian museum.

Lincolnshire, when it covered 22 acres with a six-furrow plough, using 44 gallons of paraffin in the process. The tractor weighed 10,680 lb., which is about 1,000 lb. heavier than a modern County 1164 tractor developing 116 bhp. A smaller 20-bhp version of the Marshall was added to the range in 1908.

Marshall achieved their biggest export success in Canada, where they helped to establish a place in the market by competing in the famous Winnipeg trials in 1908 and 1909. In the 1908 trials the Marshall entry came sixth out of eight entries. The next year the 30-hp model came second out of five entries in its class, and the 20-hp tractor third out of three. The top award in the Brandon trials in Manitoba in the same year was won by a Marshall tractor. Some of these early British exports have been preserved by collectors in Canada, and most of the agricultural museums in the prairie provinces have a Marshall tractor on display.

One of the last new arrivals of interest in the tractor market, before the war started in 1914, was the 'Ideal' tractor, which was manufactured in Birmingham from 1912. This was a 24-bhp tractor of considerable technical interest. It featured

23

An Ideal tractor fitted with powered mechanism for lifting implements.

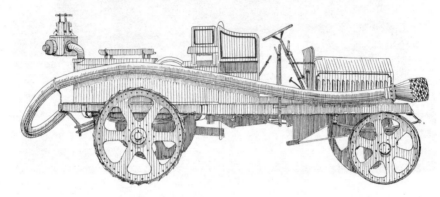

The 1909 Dennis tractor equipped with a pump for irrigation.

self-cleaning spuds on the driving wheels: the spuds retracted through the wheel rim with a cam action to scrape off the mud. There was a differential lock to the rear axle which helped to give the choice of driving to both wheels with the differential locked, both with the differential free, or to the right wheel or the left wheel only. Implements could be attached to the rear of the Ideal by means of a special frame which was part of a power-operated lifting system. The lift was worked by a chain, but was the forerunner of more modern hydraulic lift systems.

The Ideal tractor returned briefly to the market after World War I, but was too expensive and complicated for commercial success.

Failure and Success

In many ways the twenty-five years from the start of World War I to the outbreak of World War II were disappointing in terms of tractor progress in Britain. Manufacturers suffered from depressed markets, both at home and overseas, and faced strong competition from efficient American companies able to operate on a very much larger scale.

There were plenty of optimists in Britain prepared to enter the market, but few made any real impact. Some of the tractors were over-priced, some lacked reliability or were poorly designed and some were based on outdated ideas. However, there were also some which offered genuinely advanced design and deserved more commercial success than they achieved. It is a sad comment on the state of the market and state of the industry that there were no British companies producing tractors continuously from 1919 to 1939.

At a time when the industry faced such serious commercial problems, and when much of the innovation was coming from American and European manufacturers, it is difficult to find many success stories of lasting significance. Fortunately there were two developments which were an important factor in the development of Britain's modern tractor industry.

One of these developments was the decision to establish a large factory in Britain to manufacture Fordson tractors; the other was the Ferguson System, described in a later chapter, which played a part in bringing the David Brown company into tractor production and also helped to form the Massey-Ferguson organization with its important British interests.

During World War I Britain's vulnerability to U-boats seriously reduced the supply of food imported from overseas. In addition production from British farms was hampered by the loss of the horses and manpower demanded by the war effort. As it became increasingly obvious that the war would be a prolonged affair, it also became apparent that maintaining food supplies had become a critical problem.

Tractor power provided the means for British farmers to plough more land in order to increase arable production. Suddenly tractors were in demand on a scale which British manufacturers were unable to meet. Some companies, such as Marshall, had geared their production to export markets and had no tractor suitable for British conditions. There were some firms with such a small production capacity that under wartime conditions they could not expand to make any significant contribution. Fowler, Foster and others were working on contracts to produce equipment for the army. The Ivel company was fading quietly out of existence, ceasing production in 1916.

This left Saunderson as the only British company equipped for volume production of a tractor which was designed for British conditions. In order to fill the power gap, tractors had to be imported in large numbers from the United States, the only available source of supply, and this American invasion, which continued after the war and spread to other European countries, proved a disaster for British manufacturers.

The European demand helped the already vigorous American firms to expand rapidly to a size which dwarfed their European competitors. The invasion also established the names and reputations of companies such as Ford, Case and International firmly in the minds of farmers. As demand for new tractors diminished in the mid-1920s, it was the big American companies, firmly established, that were best equipped to survive – competing against British firms which had generally lost their leadership in design.

One of the development dead ends which claimed more than its fair share of resources at this time was cable ploughing. In the age of steam on the land the introduction of cable-ploughing systems had been one of the great British contributions to farm mechanization. With a cable system one steam engine or a pair can be used. The engines remain on the headland of the field being worked, and a plough or cultivators can be pulled across the field by means of a steel cable wound on to a drum or windlass attached to the engine. The system was popular in some areas of Britain, especially on heavier soils, and the engines were exported in large numbers, especially to Germany and Eastern Europe. The cable system meant that the weight of the steam engine could be kept off the land being cultivated, avoiding the problems of soil compaction and the engines becoming bogged down in difficult conditions.

The arrival of the farm tractor meant the end of cable ploughing sooner or later. The system was developed to keep the heavy steam engine, with its great power-to-weight problem, off the land. Albone, Scott, the men behind the Fordson and many others had shown that the tractor could be used for ploughing by direct traction, doing away with the need for a huge investment in cable-ploughing equipment.

Fowler of Leeds was undoubtedly the greatest name in steam-cable ploughing, and the company made several attempts to substitute internal-combustion engines for steam. The first of these 'modern' cable engines arrived from Fowler in 1913, looking remarkably like the steam engines it was supposed to replace. The resemblance was so close that the judges for the Royal Agricultural Society of England, assessing the new engine as a contender for a silver medal award, criticized the designers for their failure to break away from traditional thinking. The 1913 version was based on a previous tractor announced in 1911 and intended for stationary

Fowler 100-hp cable-ploughing engine of 1922.

work but without the winding drum for cable work. The engine was a four-cylinder petrol or paraffin unit rated at 50 bhp.

A completely new model was announced in 1920. The similarity in appearance to a traditional steam engine was gone, but the new tractor still had a horizontally mounted winding drum and was intended specifically for cable work. The 1920 version was rated at 60 bhp, and a 100-bhp model was launched in 1922. The larger tractor was designed for use with Fowler 12-bhp cable-ploughing tackle. The engine was a Ricardo design, mounted on a sprung subframe, and with a twin-cylinder donkey engine provided for starting the big engine.

Fowler's announcement of the new ploughing engines sounded less than enthusiastic. At about the time that Fordson production was reaching 100,000 tractors a year in America, and when most of the agricultural steam-engine makers were either going out of business or switching to other products, Fowler still believed, apparently, in the future of steam. The new cable-ploughing engines, it was stressed, were intended only for areas where traditional steam equipment could not be used because of unsuitable water supplies or lack of fuel.

'Under normal conditions of fuel and water supplies the makers unhesitatingly advocate the employment of steam sets, but they have equal confidence in recommending these motor outfits where there exist the abnormal circumstances already defined', announced the *Implement and Machinery Review* in 1922 after a demonstration of the 100-bhp engines.

J. & H. McLaren, a near-neighbour of Fowler in Leeds, also decided that the end of the war was a good time to launch a cable-ploughing engine, which they called a

A Walsh and Clark Victoria cable-ploughing engine.

motor windlass. The design was a break with tradition to the extent of having the windlass or winding drum mounted vertically at the rear of the tractor. The drive to the drum was through a chain instead of the more usual gear drive.

The third outstanding company producing oil engines to power cable-ploughing tackle was Walsh and Clark of Guiseley, near Leeds. This meant that Leeds was the undisputed centre for manufacturing this obsolescent equipment. The first of the Walsh and Clark cable engines were made in 1913, and in 1915 the company won a silver medal from the RASE for their 'Victoria' engine.

The most successful of the Victoria ploughing engines was produced in 1918. This version was designed to look very much like a traditional steam-ploughing engine, complete with a 'boiler' and a chimney at the front for the engine exhaust. The boiler provided the main structural strength for the machine and was made of $\frac{1}{4}$-in. boiler plate steel. The space inside was used as a fuel tank containing paraffin for more than four days' work.

For their power unit Walsh and Clark used a twin-cylinder engine with 7-in. bore and 8-in. stroke. This developed 30 to 35 bhp at 600 rpm and was mounted horizontally above its fuel tank. The engines were sold mainly to contractors, and were intended to be used in pairs, with a balance plough working to and fro between them. The cable or rope pull could be sustained at 3,500 lb., and a pair of engines working four furrows could plough up to ten acres a day.

A quite different approach to ploughing was the development of the motor plough, an idea which was fashionable in Europe and America for a few years before and after World War I. The typical motor plough design was a frame carried on two large driving wheels towards the front, and a front-mounted engine. The plough or other implements could be attached to the frame near the rear. The controls were at the back of the frame, where the operator was perched on a seat or, on some machines, walked behind on foot. Advantages were good visibility for the operator and a relatively inexpensive tractor unit.

The most popular of the British motor ploughs was the Crawley, made by the Crawley brothers who farmed in Essex, and sold as the 'Agrimotor'. The original design was developed in 1908 and was used by the brothers on their own farm. First commercial production was started in 1914, when the old-established steam engineering company, Garretts of Leiston, Suffolk, agreed to build a batch of the Agrimotors. One was shown at the Shrewsbury Royal Show in 1914, but for some reason only three of the motor ploughs were built by Garretts, and production continued in a small factory controlled by the Crawley brothers. They used American Buda four-cylinder engines developing about 30 bhp, and the Agrimotor remained in production for ten years.

Crawley motor plough of about 1919.

Most of the companies tempted to move into the market moved out again quite quickly when they discovered that demand was distinctly limited. There was a Boon motor plough, produced initially by Eagle Engineering of Warwick, and then by Ransomes of Ipswich after 1920. Petter of Yeovil and Martin of Stamford also produced small numbers.

There were at least two attempts to produce one-way motor ploughs. These had a complete plough unit of three or four furrows at each end, designed to work like a balance plough. The unit moved to and fro across a field without turning at the headlands. In each direction the plough unit at the leading or forward end was raised out of work while the trailing plough unit was working. The best known of these was the Santler from Malvern Link, Worcestershire. This had a central, upright steering column and two seats for the operator, one on each side of the steering column. The driver changed seats each time the motor plough changed direction. A Santler was entered for the 1919 tractor trials at South Carlton, Lincolnshire, but the machine failed to compete. Few were actually produced for commercial sale.

A basically similar idea, designed by Mr F. M. Waller of Boscombe, Hampshire, appears to have suffered a similar fate, with production limited to a very small number of prototype machines. Mr Waller's one-way plough went beyond the basic simplicity of a more conventional motor plough and included some fancy ideas which added complication and cost. One of these was the use of automatically self-cleaning spuds or lugs on the two driving wheels. The spuds could be adjusted to protrude beyond the rim by any required amount up to $4\frac{1}{2}$ in. The balance wheel at the rear end could be adjusted to run in line behind either of the main driving wheels, or anywhere in between their track. The ploughs were four furrow units, and the angle of lift was carefully designed so that the bodies came out of the soil one at a time to give a completely level end to each set of four furrows.

Similarly unusual, but apparently rather more popular, was the 'Rein Control' motor plough announced in 1923 by Fowler, and in production from 1924. A silver

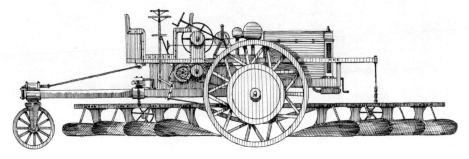

Mr Waller's design for a one-way motor plough.

medal was awarded to Fowler for the motor plough at the 1924 Royal Show. The design was intended for farmers and farm workers who were more at home controlling a team of horses. The controls for the motor plough were operated by means of reins, and the job could be done with one hand if necessary. The appropriate tug or sustained pulls on the reins controlled forward and reverse, steering to left or right, stop and start and braking. The design was imported by Fowler from Australia.

The official report of the Royal Agricultural Society judges, who watched a demonstration with the motor plough, was full of praise for the ease of control:

'A slight pull at the left or right rein steers the tractor. A steady pull on both reins stops it, and a further pull puts the reverse gear into action. The reins are easily operated by one hand, leaving the other free to attend to the implement being drawn.'

The one-way ploughs, motor ploughs and cable engines were exceptions rather than the rule, and most of the new tractors arriving on the market were more conventional in design. The most successful of the British newcomers as the war ended was the Austin. Herbert (later Sir Herbert) Austin, the highly successful motor manufacturer, announced his new tractor early in 1919. The design was almost certainly influenced by the Fordson from America, and the engine was a four-cylinder 25-bhp unit shared with one of the Austin cars in an attempt to keep the price competitive.

By August 1922 production of the Austin in Britain was claimed to total three thousand tractors. Part of the sales success had been earned by good performances at public tractor trials. The Austin had achieved a special success in France, and a factory had been opened near Paris where the tractors were assembled, using the engine and some other components imported from Birmingham. An Austin was the only British tractor to compete in the important French trials of 1919, held at St-Germaine-en-Laye. Against a strong entry of American and French tractors the Austin earned high praise and its place in the French market. The report in *Le Figaro* described the Austin as 'the acme of usefulness'.

While the French success continued, the British market declined. The 1919 price for the Austin had been £360 in Britain, and for the same money a farmer could buy three Fordsons. Although the Austin price was cut, sales were so slow that the tractor was withdrawn from the British market from 1927, leaving production and sales centred in France. In 1930, encouraged by some signs of an upturn in the tractor market, Sir Herbert Austin decided to relaunch his tractor, which was imported from France and offered to British farmers for £210. The later version of the tractor had an improved engine developing almost 35 bhp. The four-cylinder engine was claimed to be especially economical, using 0·62 lb. of paraffin per hp hour.

An Austin tractor, built in France in about 1930, with an early Rotavator attached.

Another of the great names in the early days of the motor industry in Britain was W. R. Morris, later Lord Nuffield. In the early 1920s the Morris organization was collaborating with the army in the development of a new miniature tank, known as the Morris-Martell. The tank was designed as a one-man unit, and the power unit was a 16-hp engine developed for a Morris Commercial light truck. In 1926 it was announced that Morris would be entering the farm tractor market with a tracklayer based on the new tank. A photograph of the prototype was published, but further details were not released at the time because the tank was still on the secret list. The crawler tractor failed to reach the market, and Lord Nuffield had to wait for another war to come and go before he achieved an outstanding success in the tractor market with the Nuffield Universal.

At about the time that the Morris tractor failed to appear, Armstrong Siddeley decided to move into the tractor market, and then thought better of it. The Armstrong Siddeley company was noted for quality cars at the more expensive end of the market. Their proposed tractor would almost certainly have been expensive, and would have been of outstanding interest.

The Morris tractor announced in 1926 with rear wheels for steering.

The announcement that Armstrong Siddeley were moving into tractor production came in 1927. It claimed that the car company had acquired the rights to an Italian design for a tractor in the 35–40 bhp class. The tractor would be manufactured in England and would have four-wheel drive, using four equal-sized large wheels. The design was fully articulated, with the front and rear sections linked by a drive shaft and universal joints. The front section carried the tractor engine, gearbox, controls and a seat for the driver. There was also an additional seat – rather ominously – for the mechanic. The rear section of the tractor was designed as a large load-carrying platform. It is doubtful if the tractor would have scored a commercial success in Britain if Armstrong Siddeley had put it into production. Farming opinion and economics were against unconventional, expensive tractors.

Three-wheel tractors fell rapidly from favour in Britain after the Ivel went out of production. Apart from some small economy tractors, which achieved little commercial success, it was recognized that the four-wheel configuration was more suitable for a general-purpose design. One of the exceptions was the Glasgow tractor, produced as a prototype in 1918, and in commercial production from 1919.

The Glasgow was a thoroughly unconventional tractor, designed to give extra pulling efficiency in difficult conditions. All three wheels were driven, and there was no differential mechanism in the drive to the front wheels. Instead a ratchet

An early version of the Glasgow tractor.

system was used to compensate for unequal wheel speeds as the tractor turned. The Glasgow project was controlled by a group of Scottish companies which included the well-known implement manufacturers, John Wallace and Sons of Glasgow. The tractor was produced in a former government munitions factory at Cardonald, and output was planned to reach five thousand units a year.

Although the design of the tractor appears to have been successful, the project soon ran into difficulties and was never a commercial success. A marketing contract had been agreed with an organization called British Motor Trading Corporation, and they were soon in financial difficulties. There was also the all-too-common problem of price. The retail price announced in 1919 was £450, and although this was reduced to £350 by 1922 it was still about three times the price of an imported Fordson.

Before the tractor went out of production, in about 1924, it achieved some notable successes. Export sales were relatively good, largely because of the reputation for working in adverse conditions. There was also a breakthrough in 1923 when the Dundee Territorial Association decided to buy sixteen Glasgows, after a test programme, for pulling field guns. One Glasgow with one man could replace three men and six horses for pulling eighteen-pounder guns in rough terrain. Scottish pride rose again in 1928 when a new company was formed in Glasgow to re-introduce the tractor. The company was known as Clyde Tractors Ltd, and was formed with £7,000 capital. The intention was to use more efficient production methods to achieve a price reduction to £250. Perhaps the £7,000 was insufficient; anyway the new Glasgow failed to materialize.

Another break with conventional four-wheel design was the Pick tractor of 1919, which was made at Stamford and is now mainly known for its appearance in the Lincoln tractor trials of that year. The Pick looks, in the surviving photographs, like a three-wheel tractor, with a single wheel at the front and a most unusual low-profile design. In fact the rear of the tractor was supported on a roller, instead of being on wheels, and this was driven by an enclosed 30-bhp engine with a radiator mounted at the side of the tractor. The design was presumably not a success, because the Pick Company introduced a more conventional four-wheel design in 1920.

The Glasgow tractor was powered by a 27-bhp engine, made by the American Waukesha company, and the Pick tractor of 1919 used a 30-bhp engine. This was a popular power range for tractors produced in Britain after World War I. The British Wallis tractor, introduced in 1920 by Ruston and Hornsby, started life with a 25-bhp engine, but this was increased from 1922 to 28 bhp. The British Wallis,

A Pick three-wheel tractor taking a dynamometer test at the 1919 Lincoln trial.

Alldays and Onions tractor at the 1919 Lincoln trials.

designed in America by the Wallis company, which was controlled by the J.I. Case Plow Company, remained in production until 1929.

Alldays and Onions made a small number of 30-bhp tractors at their Birmingham factory. The price of £630 in 1919 ensured commercial failure in spite of a very complete specification including fully sprung axles. Martins Cultivator Company of Stamford used a Dorman engine of 30 bhp in the four-wheel tractor they produced in 1919 as an addition to their motor ploughs. This was priced at £450, and attracted little demand.

The Vickers engineering group moved into the tractor business in 1925 with a new tractor they called the 'Aussie'. The name was apparently dropped after about three years, although the tractor remained in production until 1933. The engine of the Vickers tractor was a 30-bhp four-cylinder design, fairly typical of the mid-1920s. It started on petrol, but was equipped with a vaporizer to run on paraffin. The feature of the tractor which was claimed to be special was the patented wheel design. Each of the driving wheels was made in three separate sections. In the gaps between the sections were protruding bars, designed to remove clinging mud from the wheel rims. This was an Australian invention, perhaps surprisingly so as it was designed to cope with wet, heavy soil conditions. Not many of the tractors found a customer in Britain, but export business to Australia was more satisfactory.

One of the major developments in the industry after World War I was the formation in 1918 of a company called Agricultural and General Engineers, with a headquarters at Aldwych House, London. AGE was an ambitious attempt to join a large number of independent British companies, in the farm machinery and general engineering industries, into a large consortium. The group included some of the

36

The Aussie tractor from Vickers with rear wheels designed for wet conditions.

outstanding names in the industry, including Aveling and Porter, Burrell, Garrett, Bentall, Blackstone, Barford and Perkins, Bull Motors, Davey Paxman, Peter Brotherhood, Howard of Bedford and many others. The idea behind the scheme was sound. The group would have immense manufacturing and marketing resources with a wide product range. Instead of a large number of small- and medium-sized companies all competing against each other, the group was designed to match the strength of the big North American companies. If AGE had succeeded, Britain's performance in tractor and farm machinery development and marketing in the years between the wars might have been more impressive. In fact the group failed quite disastrously, although some of the member companies were able to survive the crash and resume operations again individually.

The AGE group had surprisingly little commercial interest in the tractor market. But although the numbers and the values involved were small, the tractors included some of the most interesting engine developments of the inter-war years.

One of the AGE tractors was the 'Peterbro', announced at the 1920 Royal Show, and manufactured by Peter Brotherhood Ltd of Peterborough, the only member of the group which was not wholly owned by AGE. The Peterbro was entered in the Lincoln tractor trials in its first year of manufacture, where it succeeded in taking second place in its class, beaten by a British Wallis. The Peterbro also took several top awards in New Zealand. A crawler version, using Roadless tracks, was announced in 1928.

A Peterbro tractor with a Ricardo engine.

The Peterbro engine was designed by Harry Ricardo, and was a petrol/paraffin design developing 30 bhp at 900 rpm. Its outstanding feature was the use of a cross-head design. Paraffin engines tend to force unburned fuel down into the crankcase where it dilutes the sump oil. The Ricardo design was intended to stop the leakage of waste paraffin past the pistons by means of tiny bleed holes. This was an elaborate and expensive means of stopping a problem which most engineers considered a nuisance to be lived with.

Like most paraffin tractors, the Peterbro was designed to use petrol to start the engine until enough heat was generated to switch to paraffin. The Blackstone tractor, manufactured by another AGE company and available in both wheeled and tracklaying versions used a special engine which could be started from cold on paraffin. Blackstone were, and still are, engine specialists, and in order to achieve a cold start on paraffin they used a fuel-injection system which introduced the paraffin as a mist into the cylinder head. The engine was also unusual in being a three-cylinder design, and in having a compressed air system for starting. In the 1920 tractor trials the Blackstone tractor earned special praise for the smoothness of its three-cylinder engine.

Both the Peterbro and the Blackstone were moderately successful by the rather dismal commercial standards of their day, and both probably deserved higher sales than they achieved. The last attempt by an AGE company to make a success in the tractor business was Garrett of Leiston. Garrett was one of the companies which survived the AGE saga, and in 1978 Garrett celebrated its 200th anniversary.

The new Garrett tractors were announced in 1929, offered with a choice of either a Blackstone or an Aveling and Porter engine. Both these were high-speed diesels, direct forerunners of the engines used in most of the world's modern tractors. In 1929 the idea of using this type of diesel engine was years ahead of its time.

Two of the tractors were entered for the World Tractor Trials held near Oxford in 1930. The Blackstone version used a four-cylinder engine, with the cylinders in two blocks of two. The compression ratio was 14 : 1, and a small air-cooled donkey engine, using petrol, was provided for starting the diesel engine. The Blackstone engine used a patented spring-injection system to introduce fuel into the cylinder. The Aveling and Porter engine was also a four-cylinder unit, but with the four cylinders in one block. The Aveling and Porter engine was an Acro design, made under licence. The Blackstone version of the tractor was rated at 35 bhp and 17·5 hp at the drawbar; the Aveling and Porter engine produced 30 bhp and 20 on the drawbar.

In the 1930 trials the smooth running of the Blackstone engine, and its clean

Blackstone crawler tractor praised for the smoothness of its engine.

A Garrett diesel tractor produced with a choice of engines in 1930–31.

exhaust, attracted favourable comment from the judges, although they also noted that it appeared to overheat and needed a bigger radiator. The Aveling and Porter engine also performed without mishap but had an incorrect governor setting, according to the judges. The trials included a record of fuel cost and consumption, which indicated that the Blackstone engine was slightly cheaper to run than the Aveling and that both could operate for half the fuel cost of a petrol/paraffin tractor of similar horsepower.

High-speed diesel engines performed so well at the trials that *Implement and Machinery Review* suggested in an editorial comment that they could be the engine of the future for farm tractors. A Garrett tractor with an Aveling and Porter engine was awarded a silver medal at the Royal Show in 1931. The judges' report, quoted in the RASE *Journal*, said:

'These four-cylinder engines, with their high speeds and even torque give that steady drive which is so essential to good ploughing. To sum up, this tractor may be described as a remarkably fine piece of English engineering workmanship that should outlast the average tractor. It was a pleasure to an engineer to look at.'

The medal, the low fuel costs and the judges' praise did little to help sales of the tractor. A publicity stunt later in 1931 also failed to attract much demand. This was

McLaren diesel tractor of 1928.

a highly successful attempt on the world non-stop ploughing record. A Garrett tractor easily broke the record by ploughing continuously for 977 hours, when the engine was accidentally throttled down too far and stopped. Although the record was already broken, the engine was immediately restarted, and a further 600 hours non-stop was completed.

Only fifteen of the Garrett tractors with wheels were produced, plus a further two fitted with Roadless tracks. The premium price, the depressed economy and the financial failure of AGE combined to finish off a remarkable tractor.

The Garrett was not the first British tractor equipped with a true diesel engine. In 1928 McLaren started using high-speed twin-cylinder diesels in a new tractor they were producing in small quantities for export. The engines were imported from Germany as a result of a link between McLaren and Mercedes-Benz.

The 1930 trials provided the first public demonstration of the new Marshall tractor with its single-cylinder two-stroke diesel engine. This was a type of engine which was popular among European manufacturers, but not in Britain, where Marshall was the only company to persist with it. The Marshall engine, with its distinctive thumping exhaust note, remained in production until 1957.

The Rushton tractor was one of several attempts between the wars to produce a British tractor to sell in large numbers. The tractor was first announced in 1928 as the 'General', but the name Rushton was used in the following year when the tractor

41

went into production. The tractor was backed by the AEC commercial vehicle company of London, and was a straightforward design, apparently based substantially on the Fordson. The Fordson link was so strong that some components fitted both tractors, and the sales director was E. Allen Webb who had previously been the tractor manager for Ford in England.

When the Rushton was announced in 1928 it was publicized in both wheeled and tracklaying versions. This was an indication of the growing interest in tracklayers among British companies. At first the Blackstone and Clayton were the significant crawler tractors manufactured in Britain, to compete against American and French imports. The Clayton was distinguished by having a steering wheel instead of steering levers.

An improved version of the Clayton was announced in 1928 with power increased to 40 hp, giving 7,250 lb. of drawbar pull operating on petrol, or 6,200 on paraffin. Clayton interests were taken over by Marshall of Gainsborough in 1930.

At about the same time some British tractors became available in Roadless tracked versions. These included the Peterbro in 1928 and the Garrett in 1931. In 1929 a half-track version of the McLaren diesel tractor became available, but there is little evidence of substantial sales.

More manufacturers moved into the crawler tractor market during the 1930s, including Fowler with a wide and rather complicated range of models. Ransomes and Rapier of Ipswich made a large tractor with a 50–65-hp Dorman-Ricardo engine, and also competed in the opposite end of the market with a small crawler tractor, using a 15-hp Ailsa Craig engine. There were also small crawler tractors from the Bristol company, powered by Douglas engines, and the much more successful MG2 tractor from Ransomes, Sims and Jefferies which, with various modifications, remained in production for about thirty years from 1936.

Half-tracks have been available from time to time, but with little success until the 1950s when there was a spasm of interest in half-track conversions. The Foster half-track was one of the early failures, too expensive to create much demand in the UK, and with possible export potential lost in the war. The tractor was described in the farming press in 1914. It was claimed to develop 4,000 lb. of drawbar pull from a 60-bhp engine. A centrifugal pump was used to circulate cooling water to the radiator, and cooling efficiency was assisted by means of a draught induced by the engine exhaust. The radiator was galvanized, and made in thirty individual sections, each of which could be removed separately for repair.

Clayton and Shuttleworth tested the market in 1917 with a large crawler tractor weighing 13 tons. This was probably designed originally for military transport purposes, but was offered as an agricultural tractor to help the ploughing campaign. The six-cylinder petrol engine was made by the National Gas Engine Company of Ashton-under-Lyne, but was described as being designed by the Munitions Mech-

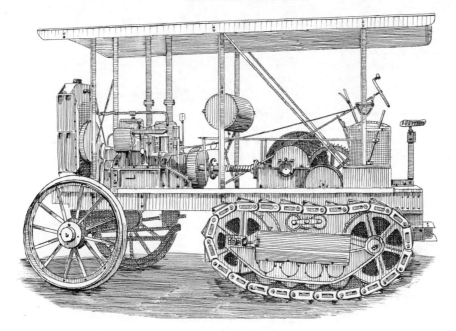

Foster Centipede tractor publicized in 1914.

anical Transport Department. The tractor was manufactured at the request of the government and was designed for crossing rough and undulating ground. The single front wheel steered the tractor, and the support arms carrying the wheel were spring-mounted to allow vertical movement. The drawbar performance was claimed to be exceptionally good, with the capacity to pull a 20-ton load.

The most ambitious of the crawler tractor projects between the wars was the Fowler Gyrotiller. This was a large tractor unit, with a single front steering wheel, and with a powered cultivator unit attached to the rear. The machine was designed by an American, N. C. Storey, to cope with the demanding conditions of sugar-cane production in the West Indies. The rotary cultivator could operate to a depth of 22 in., covering a tillage width of 10 ft. It was used for breaking up cane which was due for renewal and for working newly cleared land.

Fowler produced the first Gyrotiller in 1927 and exported it to Cuba. Most of the production machines were exported, but some in the 1930s were sold to large estates and to contractors in Britain. The first four Gyrotillers were powered by immense Ricardo petrol engines rated at 225 bhp. These were able to burn petrol at a rate of 14 gallons an hour, and a less thirsty MAN engine was substituted. There were also some smaller versions of the Gyrotiller, requiring only a 30-bhp engine.

Power steering was included in the comprehensive specification for the bigger

Clayton and Shuttleworth tractor, probably designed for army transport work.

versions of the Gyrotiller, and there was also a powered mechanism for raising the twin rotors out of work. The gearbox offered four forward ratios and two reverse, and there was a choice of two working speeds for the rotors to give control over the tilth produced.

The Gyrotiller was relatively successful and remained in production until 1939, a span of twelve years which few other tractors of the 1930s in Britain could equal. There were many reasons for the difficulties affecting tractor production, and one of these was the exceedingly effective competition of the Fordson. In a market which was seriously affected at times by economic depression and lack of confidence, the Fordson continued to sell at a price which helped to put many of its rivals out of business.

The decision to transfer the production of Fordson tractors from Ireland to Dagenham, Essex in 1933 meant that Britain for the first time had a tractor which was designed for mass production and was backed by immense resources, including a well-established overseas marketing organization. The arrival of the Fordson marked the beginning of a new phase in British tractor production and the revival of Britain's role as a major exporter of tractors.

A Fowler Gyrotiller working in Suffolk in the early 1930s.

A Fordson of 1937, produced after the move to Essex from Ireland.

An early version of the Marshall tractor made in 1936.

Different Engines, Different Fuels

Some of those who tried to influence the progress of tractor improvement were keen to use different engines and fuels. None of these attempts to introduce alternatives has made much difference to the development trend from oil and petrol to the almost universally used diesel engine. But the situation may change: as oil prices rise and world reserves dwindle we shall be forced to search for some other means of powering our tractors. Ideas which have been tried and rejected in the past may be more relevant in a different economic situation and with more advanced technology available to improve efficiency.

Dan Albone was concerned about the cost of petrol in 1905 and with characteristic determination he decided to investigate some alternatives. He was particularly interested in the use of alcohol as a fuel. Alcohol can be produced from a wide range of vegetable materials, including wood, straw and potatoes, as well as more familiarly from grain. These plant resources can be renewed annually, while our oil reserves are not replaceable. Albone thought that alcohol might be produced more cheaply than petrol, and this would help to make his tractors more economical to operate.

To compare alcohol with petrol and paraffin the Ivel tractor was set to plough, using exactly two gallons of each of the three fuels in turn. The only changes made during the test were to adjust the tractor engine to suit the fuel being used each time.

The tests showed that petrol gave the highest output, with 3 roods ($\frac{3}{4}$ acre) ploughed before the tractor ran out of fuel. The Ivel covered 2 roods and 35 poles on paraffin and 2 roods and 25 poles on alcohol.

Since Albone carried out his tests there have been other assessments of the value of alcohol as a tractor fuel. Ministry of Agriculture trials in Britain during World War II produced an estimate that the alcohol from two acres of potatoes would power a 20-hp tractor for a year of normal farm use. An American research team working in 1910 found that one bushel of potatoes would produce 1·5 US gallons of alcohol. They calculated that the alcohol from an acre of potatoes – 600 gallons – would provide the power for a tractor to plough 200 acres. To plough the same acreage with horses would require the grass, hay and oats from 40 acres to provide feed for the eight horses which would be needed, the team claimed.

Electric power for tractors has never attracted as much interest in Britain as in France, Germany and Russia. Cable ploughing outfits powered by electricity were working in Germany in the early 1890s, and as recently as 1951 the British farming press carried detailed reports of what was claimed to be the widespread development of cultivation systems powered by electricity in the Soviet Union.

A more successful attempt to use electricity in the field was reported in 1925 from Scotland, where Major Andrew McDowall of Drem, East Lothian, demonstrated an electric ploughing tractor he had designed. The tractor was a three-wheel design, with one-way ploughs mounted at each end. It was propelled by a 12·5 hp electric motor, taking current through a cable from an overhead power line.

The tractor was developed to work to and fro across a field, with each plough alternately lifted or in work. Power from the overhead line was carried direct to a mobile stand which was positioned half-way along the line to be ploughed. A 200-yard cable from the tractor engine was plugged into the mobile stand, and was automatically wound on to a drum carried on the tractor as it approached the stand, and then wound out again as the tractor continued to the headland. The 200-yard cable allowed a total length of 400 yards to be ploughed.

Major McDowall claimed that electricity was so much cheaper than petrol or paraffin that his equipment cut the cost of ploughing by 50 per cent. The next stage of development he planned was to automate the tractor with fail-safe devices front and rear, so that the tractor would stop if it hit an obstruction, and with a self-steering mechanism. Four years later, in 1929, the Major asked the Royal Highland and Agricultural Society to support a claim he was making for a grant of £3,000 from the Department of Agriculture for Scotland, to cover the cost of further development of his tractor. His request was turned down, although the RHAS expressed some interest in his idea.

It was the steam engine which provided the most determined challenge. Steam had contributed significantly to the commercial success and prosperity which Great Britain enjoyed during the nineteenth century.

Some of the old steam-engine firms continued to ignore the arrival and success of the tractor, and persisted with their traditional product lines until the market disappeared. There were also some which stayed with steam but tried to modernize and adapt their engines to compete more effectively with the tractor. Some of these steam tractors were a complete break with tradition, and British and American firms were among those producing the most efficient agricultural steam engines ever built. They all failed because the original thinking and improved technology which produced them arrived too late: the internal-combustion engine had taken a complete hold of the market.

One of the first of the new-look steam tractors was the 'Suffolk Punch' manufactured by Garrett of Leiston, Suffolk, one of the oldest companies in agricultural engineering. Work on the original design for the Suffolk Punch began in 1915, and the aim was to produce a multi-purpose power unit which could be the basis for an agricultural tractor and also for a waggon for road haulage.

A feature of the Suffolk Punch was that it had been designed for direct traction, for which the traditional type of agricultural engines was unsuitable. The specification included a superheater to raise the boiler temperature to 600°F. The boiler was a horizontal type, but with the firebox end towards the front of the tractor and the chimney at the rear. The driver was positioned right at the front, with his back to the firebox, and with controls which included a vertical steering column with Ackermann steering mechanism. The axles were sprung, and the engine was a double-crank compound type, rated at 40 bhp. Power was delivered to the rear driving wheels by means of a two-inch pitch roller chain.

Only eight of the Suffolk Punch tractors, or 'Agrimotors' as they were called, were built by Garrett, who then abandoned the project in the face of low-priced competition from the new tractors – such as the Fordson – which were arriving on the market in increasing numbers.

Another attempt to revolutionize the agricultural steam engine was launched at about the same time by the Summerscales company of Keighley, Yorkshire. Summerscales was a medium-sized general engineering firm founded in 1850. Their attempt to enter the tractor market as World War I was drawing to a close was part of a general plan to diversify and strengthen the company, which was in financial difficulties. In fact the tractor and the company both failed commercially, and the financial difficulties were explained in 1920 when Summerscales was wound up. Apparently a decision had been made to invest heavily in equipment for making ammunition for the British army, and the war had ended before the investment could be recouped.

A Summerscales tractor took part in the 1919 tractor trials held at South Carlton, Lincoln, and this was one of only two which are thought to have been built. The boiler was a mid-mounted vertical design, producing steam for a four-cylinder engine with the cylinders in a V-formation. The steam was superheated, and the engine was rated at 25 bhp. Drive to the rear wheels was by a roller chain, and there was a single front wheel – probably a disadvantage in a tractor weighing four tons. The price quoted in 1919 was £600, which did not compete with the price of the Fordson imported from America.

In spite of the disappointments experienced by Garrett and Summerscales, other manufacturers were ready to interest farmers in steam power. The Sentinel Company of Shrewsbury, Shropshire, produced their first steam tractor in 1924. It was based on the design of a steam waggon for road transport work which the company produced with outstanding success.

The Super-Sentinel steam waggon was announced in 1923, and was probably one of the finest steam vehicles ever made. The following year a rough-terrain version of the Super-Sentinel was announced, designed to operate in areas with poor or

Sentinel-Roadless steam tractor made in 1924.

non-existent road surfaces. It was this machine which encouraged the company to design a further development of the Super-Sentinel specifically for agricultural and forestry work, including ploughing by direct traction.

Much of the tractor, which was launched in 1924, was still the original Super-Sentinel design, including the front-mounted vertical boiler and the Ackermann steering for the front wheels. The main differences between the first rough-terrain design and the agricultural model were that larger rear wheels were fitted for farm conditions, and a lower ratio was used for the single forward gear. Later an optional second forward ratio was offered. A peculiarity of the dual-speed model was that to select either ratio the tractor had to be stationary, and there was no safety device to prevent both gear ratios being selected simultaneously.

A successful version of the tractor was fitted with Roadless tracks in place of the rear driving wheels. This increased the price – to £1,250 instead of £880 for the four-wheeled version – but also increased the usefulness of the tractor in East Africa, where most were sold. Production of the Sentinel-Roadless amounted to approximately seventeen, each weighing about ten tons.

The Sentinel tractors were available with an oil-fired conversion, and with a superheater to raise the working temperature of the steam to between 600° and

650°F. The boiler was designed to operate at 230 psi. Work rates of 20 acres a day were reported by some colonial owners using the Sentinel-Roadless with disc ploughs or pulling eight furrows.

Pleased with the success of their tractor, the Sentinel company launched an even more powerful and more expensive version in 1927, which they called the 'Rhinoceros' – presumably with their African customers in mind. The boiler pressure was raised to 275 psi, and the output was rated at 86 bhp. The price quoted in 1927 was £1,800, and a total of eight of the tractors were built, most of them exported to Africa.

The Rhinoceros was equipped with twin water tanks, one at the front and the other at the rear, with a total capacity of 260 gallons. In full steam the boiler was capable of evaporating 2,300 lb. of water an hour, and produced up to 22,400 lb. of drawbar pull when starting from rest. The tractor was claimed to pull a load of 75 tons on a hard level road surface. At about the same time the Sentinel company was developing a new heavy-duty agricultural tractor to be fitted with a twin-cylinder McLaren-Benz diesel engine, their first departure from steam. The diesel tractor project was abandoned before production had begun.

The tractor which overtook the Rhinoceros in the steam power race arrived in 1931 from Foden of Sandbach, Cheshire. It was their new steam tractor, aimed at the African and South American farming markets and claimed to be 'the most powerful and economical agricultural tractor in the world'. The general layout of the Foden was similar to that of a steam-traction engine, with a horizontal boiler which was coal fired and rated at 250 psi with a superheater to raise the steam temperature to 600°F. The engine was double-acting with cylinders of 5-in. bore and 7-in. stroke, and power output was rated at 100 bhp.

The Foden 100-hp steam tractor designed mainly for export.

The Foden 'AgriTractor', as it was called, was probably the last attempt by a British manufacturer to develop a steam tractor, and it was a commercial disaster. A considerable promotional campaign in likely export markets found only two or three customers, and Foden then abandoned their venture into the tractor business.

When normal fuel supplies have failed in the past, some farmers have turned to methane gas as an alternative. Methane has the considerable attraction that it can be produced on the farm from raw materials which are easily obtained and renewable.

The gas is produced by a fermentation process. Suitable equipment for the process and for storing the gas could be installed on a farm scale. The raw materials could be waste by-products such as farmyard manure and slurry, or cereal straw. It has been calculated that the manure produced by all the housed farm livestock in Britain could be used to yield methane with an energy value equivalent to about 30 million gallons of petrol.

During World War II large numbers of tractors, as well as cars and other vehicles, were converted to run on methane and other gases, especially in France. An

A Fordson tractor converted by the NIAE for methane gas trials.

estate in Surrey was equipped with its own methane-production equipment in 1922 to provide gas for use in the main house and estate cottages, and to run a stationary engine. This equipment was still in use, with some modifications, in the early 1950s. The National Institute of Agricultural Engineering (NIAE), then at Askham Bryan in Yorkshire, carried out a series of trials with Fordson tractors converted to run on methane and suction gas soon after the war, when fuel supplies were still difficult.

As supplies of coal and oil became plentiful again, the research and the gas conversions were abandoned. Farming with gas as a fuel has its disadvantages, and methane cannot compete with cheap, plentiful petrol and diesel oil. One of the problems is a significant loss of power when a tractor is converted to run on methane instead of petrol or paraffin. This can amount to a 20 to 50 per cent reduction, according to French research. The French trials, using suction gas from charcoal, suggested that even a 50 per cent power loss could be tolerated, because the gas produced on the farm was up to 75 per cent cheaper than petrol when the trials were carried out in 1924. The extra labour cost for a tractor driver working at half-speed seemed of little consequence more than half a century ago.

Another disadvantage is loss of convenience. The easy way to obtain tractor fuel is to telephone the local supplier, and complain if he takes more than a few days to deliver. Fuel costs will no doubt rise considerably before many farmers exchange a good delivery service for a methane plant which requires frequent attention, plus the inconvenience of refilling bulky fuel containers on the tractor.

Before a fuel crisis perhaps forces us to switch to methane or some other home-produced energy source we must hope that technology can overcome some of the disadvantages. It may be possible to automate small digesters to make them less time-consuming to operate. Another step forward would be a convenient method of filling the fuel container on a tractor with gas at sufficiently high pressure so that a smaller container could be used to give a full day's work. The biggest problem to be solved, at the moment, is the loss of efficiency in the digestion process as the temperature falls. The average winter temperature in Britain is too low for the process to work properly, and this is why there has been more commercial interest in it recently in warmer countries, such as South Africa.

The Ferguson System

Most of the tractors in use throughout the world have features developed originally by Harry Ferguson. He and his team made an immense contribution to improving the efficiency of farm mechanization, and he also established a highly successful commercial empire.

The success story had a modest beginning on a farm in what is now Northern Ireland, where Harry Ferguson was born on 4 November 1884. The family farm at Growell, County Down, was about a hundred acres, larger than the average in the area but too small to provide more than a fairly simple home.

Living and working on the family farm probably helped Harry Ferguson in his later career, although the toil and routine of a farm worked with horses and man-power were not to his liking. Another factor in his childhood was the austerely religious way of life which his father imposed on the family, and against which Harry increasingly rebelled. Before his eighteenth birthday he decided to leave home to find a way of life away from the farm. He was offered an opportunity to join an elder brother in a garage business he was building up in Belfast. Harry welcomed the idea, which suited his interest in cars and engines.

The garage prospered and built up a good reputation, helped considerably by Harry's talent for tuning the unreliable engines in use at that time. The business also benefited from his racing success with motor cycles, which was exploited to the full for publicity for the garage.

Another of his early talents, and an outlet for his mechanical aptitude, was air-craft design. When Ferguson first joined his brother's business in 1902 they were operating on a tight budget. By 1908 there was sufficient money to finance Ferguson's ambition to design and fly his own aircraft. After numerous failures he made his first flight in December 1909. This was the first time an aircraft had flown in Ireland, and Harry Ferguson was the first person in Britain to design, build and fly his own aircraft. More flights followed, and he might have considered making his career with aircraft, but in 1911 his energies were diverted to making a success of a garage business of his own after breaking away from his brother.

The new garage business included the agency for Vauxhall cars, and Harry Ferguson achieved a good deal of local publicity with Vauxhall on the racing track. Much later, in the 1920s, he became a leading figure in the campaign to establish a major motor sport event in Northern Ireland. His influence and energy played an important part in starting the famous Ulster Tourist Trophy races in 1928. These

Harry Ferguson.

attracted some of the most famous cars and drivers in the world to Northern Ireland, drawn by the prestige of the Ards circuit.

Harry Ferguson's interest in farm mechanization developed during World War I. Tractors had suddenly become important as the key to increasing food production, and the government launched a ploughing campaign to turn large areas of pasture into more productive arable land. Imported American tractors played a major part in the campaign, and the Ferguson garage held the agency for the 'Overtime'. This was the British name for one of the American imports, a sturdily built and reasonably reliable tractor which became popular in Britain.

Harry Ferguson relied on demonstrations to overcome some of the sales resistance to the tractor, and these were usually carried out with a plough. His activities with the Overtime were noticed by the Irish Board of Agriculture, and in 1917 he was asked to help to improve the standard of tractor operation on Irish farms. Many of the tractors were difficult to use, especially with inexperienced operators and with implements designed originally for use with horses. This work took Ferguson, and a talented engineer from his garage who was called William Sands, to many farms where tractors were in use, where they explained and demonstrated how to get the best from the equipment.

What he saw on his travels convinced Ferguson that there must be a more efficient way to use tractor power in the field, and he set to work on the problem. His first approach was to design a plough with a low draft requirement, light in weight, and with an ingenious arrangement of springs to help the tractor driver to raise it out of work. The plough was designed to hitch so closely behind the tractor that depth wheels would not be needed. The plough hitch point was beneath the tractor and ahead of the rear axle, so that the pull on the plough helped to stabilize the tractor and improve the grip of the rear wheels.

With so much variation in tractor dimensions, drawbar positions and power, Ferguson decided to make the plough suitable for one make of tractor. He chose the 'Eros', which was a tractor conversion for the Ford Model T car. The kit was made by the E. G. Staude Manufacturing Company of St Paul, Minnesota, and was one of several conversion kits made for the Model T.

The plough went into production, to be sold as the 'Belfast'. It attracted some interest and favourable comments on its unusual design, but it failed commercially because Henry Ford's new Fordson tractors were arriving in large numbers from America. The Fordson put conversions of the Eros type out of business, and with them went the sales potential for the Ferguson plough.

A new plough was designed specifically for the Fordson. This version was also designed to work without a depth wheel. The hitch arrangement consisted of two sets of struts, arranged to keep the plough in work and also to transfer shock loadings on the plough as a downward force at the front of the tractor to prevent the danger of overturning. The struts were the forerunner of a three-point linkage, and were known as the Duplex hitch.

The new plough took Ferguson to the United States where he formed a joint company with the Sherman brothers of Evansville, New York State, to manufacture an improved version of the plough for the American market. Meanwhile development work on the hitch design continued, and in 1928 a hydraulic system for operating the draft control principle was tried as an alternative to a mechanical design. In the same year it was announced that production of Fordson tractors would be transferred from America to Ireland, and this led to the collapse of the Ferguson-Sherman plough business in America.

Fortunately Ferguson was strong enough financially to survive the American set-back, and to continue his development work. The Duplex hitch was replaced by a series of experimental arrangements of converging links and systems of three linkage points.

With the Ferguson System on the way towards completion efforts were made to find a manufacturer to put it into production. In his book *Harry Ferguson, Inventor and Pioneer* (published by John Murray) Colin Fraser cites some unexpected names among the companies which were possible partners for Ferguson at that stage. Some of the shipbuilding companies in Northern Ireland were approached, because they had the spare capacity at the time and because of Harry Ferguson's hope of building the tractor in Ulster. In America Allis-Chalmers took out an option on some of Ferguson's patents, and in Britain the Rover car company looked closely at the Ferguson System as a possible way into the tractor market. Ransomes and Rapier of Ipswich and the Rushton tractor company also showed interest, as did Morris Motors of Oxford.

As more companies proved unwilling or unable to find the resources to put the Ferguson ideas into production, it became clear that the advantages would have to be demonstrated. In order to carry out a demonstration, a tractor incorporating the Ferguson patents was required together with some suitable implements. A completely new tractor was designed and built by the team, and was finished in 1933 with a coat of gloss black paint. It became known as the 'Black Tractor', and deserves its place in the Science Museum, London, as the tractor which brought the Ferguson System into the world.

The Black Tractor was powered by a Hercules engine from America, producing 18 hp. At the rear of the tractor was an almost complete Ferguson System, with the linkage in its now familiar arrangement of a single upper point and two lower arms. The single top arm of the linkage was used, after some trial and error, to actuate the hydraulics of the draft control system using compression forces coming up from the implement.

Some of the components for the Black Tractor, including the gears, were sup-

The Ferguson Black Tractor now preserved in the Science Museum, London.

plied by the David Brown company of Huddersfield. This established a contact between Harry Ferguson and the company, which led in 1935 to the manufacturing agreement Ferguson had been seeking. The manufacturer was David Brown Tractors, which occupied space in premises belonging to the gear company. The marketing company was controlled by Harry Ferguson and his backers. The product was a Ferguson tractor, often referred to as the 'Ferguson-Brown'.

New Ferguson tractors started to arrive at the end of the Huddersfield production line in 1936. Demonstrations, organized with the precise Ferguson eye for detail, were arranged to show the advantages of the tractor, but sales were slow. There was resistance to the price, which at £224 was almost twice the cost of a Fordson. To buy a Ferguson meant additional expense for the special implements required, whereas a Fordson would probably suit the existing equipment on the farm.

The paint finish for the tractors was grey, and this remained the standard colour for Ferguson tractors until the Massey-Harris red took over. Styling was obviously based on the Black Tractor, but in the production model the engines were an 18–20-hp Coventry Climax E in the first five hundred tractors, and a David Brown engine of 2010 cc for the rest of the production run. The gearbox had three forward ratios and a reverse.

In order to encourage sales a special training school was set up by Harry Fer-

Ferguson tractors in the David Brown assembly plant in 1937.

guson. The aim was to improve the standards of servicing and operation of the tractors to ensure that their performance was up to standard.

As the sales position remained disappointing relations between the Ferguson team and those at David Brown began to deteriorate. The tractor sold reasonably well in parts of Scandinavia and in Scotland, where its special advantages were most useful, but in spite of this stocks of unsold tractors began to accumulate at the factory, bringing cash-flow problems. There was some argument about teething troubles in the design and manufacture of the tractors, and disagreement between David Brown and Ferguson over changing the design.

David Brown (now Sir David) believed a more powerful engine and a fourth forward ratio in the gearbox would help to make the tractor more saleable. Ferguson, who found it hard to work harmoniously with any of his business partners, insisted that his original design was right and refused to agree to any changes.

Relations became even more strained after David Brown announced that he was briefing a design team from his company staff to work out the details for a tractor incorporating the improvements he considered necessary. Then in 1938 Harry Ferguson arranged to send a Ferguson tractor and implements to the United States. Through Eber Sherman, his former business partner in America, contact had been re-established with Henry Ford, and the tractor on its way to America was to be demonstrated to him.

The demonstration took place in autumn 1938, apparently without the knowledge of David Brown. The 'Ferguson-Brown' showed its advantages, under the supervision of Harry Ferguson, and Ford was suitably impressed. That same day Ferguson and Ford shook hands on an agreement to work together to produce a new Ferguson System tractor. The agreement, involving millions of dollars of Ford money and the patents which were Ferguson's life's work, was never witnessed or written down.

Meanwhile back in Yorkshire the partnership between David Brown and Harry Ferguson ended in January 1939 in complete discord. Fortunately for the David Brown company – and for the future of the British tractor industry – the David Brown plans for a new tractor with more power were already well advanced, and in spite of some disappointments with the old Ferguson tractor, David Brown decided to stay in the tractor business after the break with Ferguson.

The new David Brown tractor – the first to carry the company's name – was launched in July 1939. The new Ford tractor, resulting from the agreement with Ferguson, was launched at the end of June 1939 in America. The Ford tractor was called the 9N, the David Brown was the VAK 1.

Although the 9N was a Ferguson System tractor it was completely different in design and styling to the Ferguson-Brown. It is an extraordinary tribute to Henry Ford, and the resources he controlled, that only eight months passed between the first Ferguson-Brown demonstration in America, and the huge public demonstration of the new 9N.

Harry Ferguson and Henry Ford both shared a conviction that the new tractor had a vital role to play in the mechanization of world farming. This was to be a major contribution to improving the efficiency of food production and raising the living standards of farmers. To help achieve this objective the price of the 9N was to be kept within the financial reach of as many farmers as possible.

Production of the 9N tractor continued through the war years, with output rising in several years above the 40,000 mark, and the tractor made a big impact in the American market. Meanwhile in Britain Ferguson was again making determined efforts to find a manufacturer to make a tractor to his design. He had expected that the British end of the Ford empire would be ready to adopt his ideas, but this did not materialize. The Fordson tractor was in volume production and was well accepted by British farmers. It would have been difficult to introduce a new model under wartime conditions, and there is little evidence that the directors of Ford in England were enthusiastic about Ferguson or his ideas.

Since Ford – his first hope for a British partner – was a non-starter as far as any deal was concerned, Harry Ferguson looked for an alternative. His search ended soon after the war when he reached an agreement with Sir John Black, then managing director of the Standard Motor Company, to manufacture a new tractor in Coventry. The Standard Company, now absorbed anonymously into the British Leyland organization, had spare factory capacity available at Banner Lane where the company had been producing aircraft engines for the war effort.

Ferguson's determination and high-level contacts won through in the campaign to obtain steel allocations for the new tractor, and also the release of scarce dollars to pay for an American engine until a British alternative became available.

By the end of 1946 the first of the famous TE20 Fergusons were emerging from the Banner Lane production line. They were powered by a Continental engine imported from America, which was an overhead-valve design developing 23·9 belt hp as a maximum, but rated at 20·3 bhp. The engine capacity was 1966 cc, with four cylinders, operating on petrol. Importing this engine had been a controversial move, given official approval only because the expenditure of dollars involved would help to get the Ferguson tractor into production and into export markets more quickly. The replacement engine from Standard became available in 1948, and this was also a four-cylinder design, developing about the same power as the Continental, but with slightly smaller capacity at 1850 cc.

Much of the design of the new tractor was obviously influenced by the Ford 9N, but the TE20 used a gearbox with four forward ratios, instead of the three Ferguson had chosen for the 9N and the old Ferguson-Brown tractors.

The launch of the TE tractor in England was good news for Ferguson, but 1946 also brought him bad news from America, where his agreement with Ford over the 9N tractor was beginning to crumble. Henry Ford II, now in charge of the huge company, was not satisfied with the old marketing arrangement which allowed Ferguson control over the tractor marketing organization. By the end of 1946 the discontent on the Ford side was becoming obvious, and it led to a new company being formed by Ford to take over the tractor marketing. This new arrangement was to become effective in July 1947, when the Ferguson organization in America would no longer have a tractor to handle.

July 1947 dealt a further blow to the Ferguson interests when Ford announced a new tractor, the 8N, to replace the 9N which went out of production. The new tractor, designed by Ford engineers, was equipped with a full Ferguson System linkage and hydraulics. The tractor was based on its predecessor, with similar styling, but the old three-speed gearbox was replaced with a new box giving four ratios and a reverse.

One of the results of these developments was a crisis for Ferguson's American company, as the dealers they once supplied signed up with the new Ford marketing company, called Dearborn Motors. Efforts were made to find a factory and some financial backing to enable Ferguson System tractors to go into production in America under Ferguson's control. Eventually production of an American version of the TE20 was arranged in a factory in Detroit. This tractor, known as the TO20, used Continental engines similar to those which had been imported to start the TE production line in Coventry. Later an improved version, the TO30, was introduced from the Detroit factory, and sales topped 30,000 units in 1951 and in 1952. In both years total production of Fergusons in Coventry and Detroit exceeded 100,000 tractors.

Another result of the action by the Ford Company was the famous lawsuit in which Harry Ferguson and his companies claimed damages from the Ford Motor Company amounting to $251,000,000. This sum was later increased by a further $90,000,000. Part of the claim was for compensation for the alleged damage to Ferguson's interest by the changed distribution arrangements. There was also a claim for compensation for what Ferguson alleged was the unauthorized use of his patents in the new 8N tractor.

A Ferguson tractor equipped for the 1957 Antarctic expedition.

The legal proceedings were complicated by the informal nature of the original handshake agreement on which the tractor production and marketing arrangements had been based. They were also prolonged by Ferguson's unwillingness to accept an out-of-court settlement. The proceedings dragged on for more than four years, during which Ferguson tended to gain public sympathy as the (relatively) small businessman confronted by the vast Ford organization. At the same time there was strong demand for tractors, and Ferguson's business interests appeared to be prospering as production of the 'Fergie' increased. This was hardly helpful to his claim to have been so seriously damaged by the formation of Dearborn Motors.

Harry Ferguson eventually agreed to a settlement in 1952. He accepted a judgment for payment of $9·25 million by Ford to cover the royalties on his patents, plus an agreement by Ford to stop using some patented features on the 8N. The claim for compensation for damages to his company by the changed tractor marketing arrangements was dismissed. The verdict was not the great victory Ferguson had hoped for, although it was popularly interpreted as such. The net amount received by Ferguson and his interests was fairly modest after legal costs, amounting to more than $3 million, had been paid.

The former farm boy from Ireland now owned a Rolls-Royce and a magnificent country house in the Cotswolds. The Ferguson System was making a significant contribution to the efficiency of farm mechanization, and the advantages of the ideas Ferguson had worked so hard to develop and promote were no longer in doubt. The rewards Ferguson earned from the commercial development of his ideas were considerable, but they might easily have been much greater. One of the limiting factors on the profits earned by his companies was Harry Ferguson's policy of holding down his selling prices in a personal campaign to check inflation. This was a cause he believed in and promoted with great enthusiasm and little support. He applied the principles to his own business and he encouraged others to follow his example. He paid for advertising space in the national press to explain his views, and wrote numerous letters to politicians in an attempt to obtain government support. He also provided a substantial cash prize to be awarded for the best contribution to putting his ideas into practice.

The Ferguson formula for economic reform was to introduce a policy of price reduction. This was the key to breaking the spirals of cost and wage inflation. With prices no longer increasing, wages could be pegged so that increased income could be achieved only through promotion or higher productivity. Profits must also be restricted, with excess profits being invested for improved efficiency.

In the years around 1950, when the Ferguson economic theories were promoted most vigorously, Britain was struggling with an annual inflation rate of about 5 per cent. Ferguson predicted that the rate of inflation would increase disastrously if it was not brought under control by the method he advocated, and the result would be serious unemployment and encouragement to the spread of Communism. Since then, as successive governments have failed to cope with inflation, Ferguson's credo has not been disproved, and his assessment appears increasingly relevant.

Soon after the lawsuit with the Ford organization had ended Harry Ferguson entered the final phase of his career as a leading figure in the world tractor industry. The new venture appears to have developed after his American company, based in Detroit, faced increasing financial difficulties. The answer to the American problem, Ferguson decided, would be to interest Massey-Harris in taking over the Detroit company. Massey-Harris, a Canadian-based company, seemed a logical suitor for several reasons. There was already an established contact between Massey-Harris and Ferguson, and although the Canadian company had a highly successful product range in farm machinery, this was not matched by their tractors. Massey-Harris had the resources to buy and revitalize the American Ferguson business, and the Ferguson business would give them the successful tractors they lacked.

This approach, in 1953, failed. The Massey-Harris board decided that the ailing Detroit company on its own could be of only limited value to their international business interests. However, it made both sides conscious of the possibilities of some more fundamental link, and an opportunity to discuss this occurred later in 1953 when the Massey-Harris president, James Duncan, visited England to see a demonstration of a new prototype Ferguson tractor.

During this visit Ferguson began the negotiations which provided the basis for a link between the two companies. Agreement was reached for Massey-Harris to buy Ferguson's interest in his companies for $16 million. Payment was in Massey-Harris shares, and Harry Ferguson became the largest single shareholder in the new group. The agreement provided Ferguson with the honorary position of chairman of the new company, with some executive responsibility for major engineering decisions. It was also agreed that the Ferguson name would be perpetuated in the new company, which was to be called Massey-Harris-Ferguson. This was later shortened to Massey-Ferguson.

The new organization was a logical union of different strengths, but a difficult settling-in period was to elapse before the benefits began to show. During this time important and difficult decisions had to be taken, and as usual Harry Ferguson had strong views which he promoted vigorously. The situation began to deteriorate, and there was probably a serious danger that a rift between Harry Ferguson and the M-H-F board might have soured the already difficult relationships between former Ferguson employees and their Massey-Harris colleagues. Matters came to a head when Harry Ferguson threatened to resign and to sell his shares in the company. The M-H-F board decided to accept his resignation and to arrange for his shares to be purchased.

In practical terms, that was the end of Harry Ferguson as an active force in the tractor world. But it was certainly not the end of Ferguson as a man of energy and ambition. He turned his attention and his resources to development work for the motor industry. In his company, Harry Ferguson Research Ltd, he had built a strong team to carry out engineering research.

At one stage he seriously considered making a return to tractor development. He believed that the FE35 – the first new tractor launched by Massey-Ferguson – had broken away from his original philosophy of simplicity and economical, functional design. He talked of a new Ferguson tractor which would be based on his ideals, and would continue his ambition to bring the Ferguson System to as many farmers as possible.

The new tractor never materialized, and the response from the car industry to the developments he had produced was disappointing. Harry Ferguson was facing problems of failing health and he could no longer rely on the vigour and forcefulness which helped him to build a commercial empire.

Harry Ferguson died in October 1960. His record of achievement in the tractor industry would be hard to equal, and the evidence of his achievements is to be found on most of the world's tractors.

A Time for Expansion

World War II brought increased prosperity to British farming and a big demand for tractors. The position was comparable to that in World War I. Enemy action was disrupting food supplies from abroad, and the problem of increasing production from British farms became a high priority. This demanded mechanization, and the extra tractor power had to be shipped across the Atlantic.

Among British manufacturers only Ford had the capacity to produce tractors in large numbers. As the likelihood of war had increased, Ford had stepped up their production, anticipating the demand for tractors which would follow the declaration. The gamble, which had official support from the British government, helped British agriculture at a critical time, and it also helped Ford to take a remarkably dominant hold of the market with more than 90 per cent of wartime production.

As the war ended it became obvious that demand for tractors would continue at a high level, both in Britain and in other agriculturally developed countries. Britain was in a strong position to supply many of the tractors which would be required, in spite of the problems involved in recovering its economic and industrial structure from wartime damage.

Britain's strongest rivals in the world market were the big American manufacturers, most of which were already well established in export marketing. For these companies an important problem was the strength of the American dollar, which meant that many of their potential customers were unable to afford the tractors they wanted. Britain was one of the countries which lacked the dollars to import tractors, and the British government was anxious to increase production in the United Kingdom, not only to reduce the necessity for imports but also to rebuild the country's export business.

The favourable prospects for manufacturing in Britain brought important developments. Some of the big North American companies were encouraged to establish assembly or manufacturing facilities in Britain, attracted both by the growing domestic market and by the export possibilities for tractors from Britain. Besides the transatlantic invasion there was also a small avalanche of new British companies on the market. The third development was that all the established British manufacturers brought out new models soon after the war to improve their competitive position.

Among the early arrivals from America was Minneapolis-Moline, which launched the UDS tractor in 1948. This was one of several unsuccessful attempts to interest

Minneapolis–Moline UDS tractor 1948.

British farmers in a relatively large tractor before the market was ready for it. An interesting feature of the UDS was that it arrived on the market with a choice of two diesel engines. The smaller engine was a four-cylinder Dorman rated at 46 bhp and claimed to develop 6,000 lb. of drawbar pull. The alternative was a 65-bhp Meadows unit which raised the drawbar pull to an advertised 7,000 lb. Both versions used a gearbox with five forward ratios, controlled by a hand clutch which was said to be easier than a foot clutch to operate.

Standard equipment on the UDS included a power-take-off shaft and a pulley for belt work. The driving seat could be adjusted sideways to give an off–centre operating position when required. Electric starting was also standard, but this was backed up by hand starting with a decompression device. The UDS selling price in 1949 was £1,200 plus delivery for the more expensive Meadows version. This was nearly £1,000 more than the price of a basic Fordson, and was more than the market would stand. Minneapolis–Moline and the UDS made only a brief appearance in Britain.

Allis–Chalmers started their tractor operations in Britain from an assembly plant near Southampton in 1948, but later moved to Essendine, Lincolnshire. British

Allis-Chalmers D270 tractor, an upgraded version of the famous model B.

farmers were offered the model 'B' tractor, originally produced in America in 1937 and familiar in Britain as a wartime import. The tractor was launched with a four-cylinder petrol/paraffin engine rated at 16·3 bhp, but this was upgraded in 1952 to develop 19·5 bhp at 1,500 rpm. In 1954 a diesel version was announced with a Perkins P3 engine. A completely restyled version of the 'B', called the D270 was launched in 1956, followed by a diesel version using the P3 engine and known as the D272. Although their tractors achieved only a small share of the British market, after some initial success with the 'B', Allis-Chalmers made considerable impact with farm machinery.

International Harvester has been one of the most consistently successful of the American farm machinery companies in Britain since 1851 when a McCormick reaper won a gold medal awarded by the RASE. IH tractor production started in Britain in the company's Doncaster factory in 1949. The first British-built tractors were the BM series, based on the popular model 'M' tractors originally launched in America in 1939, and familiar in Britain as an import during the war.

The BM was powered by a four-cylinder petrol/paraffin engine of 39 bhp. In 1952 the BMD, a diesel-engined version, was added to the range, and both were upgraded with more horsepower the following year as the Super BM and Super

International Super BMD tractor, 1954.

BMD. Also in 1953 the first International crawler tractors manufactured in Britain were launched, as the BT6 and BTD6. These used basically the same engines as the BM and BMD respectively.

The first IH tractor designed and built in Britain was the B-250, announced in 1956. This was a diesel tractor, rated at 30 bhp, and was manufactured in a factory at Bradford, purchased from the former Jowett car company. A larger model, the B-275, was announced in 1958 and earned considerable publicity as the tractor used by the British team which achieved first and third places in the 1960 World Ploughing Championships in Rome. The B-250 and B-275 tractors both firmly established the British IH company as a major exporter, notably to the United States and Canada. The 40-hp B-414, announced in 1961, added to the export success and helped the British IH company to win its first Queen's Award for export achievement in 1966. The B-414 was also used by a World Ploughing Championship winner in 1962, when the trophy was awarded to a Norwegian.

Massey-Harris produced their first batch of tractors in Britain in 1948 from their Barton Dock Road factory in Manchester. This factory was already producing a

Massey-Harris 744D tractor in unusual half-track form.

range of farm machinery, and the tractor operation was started in order to strengthen the company's position in the British market. Another reason was the prospect of export sales to countries which, for various reasons, it was difficult to supply from North America.

Before launching the tractor, the 744D, Massey-Harris attempted to negotiate with Nuffield, and later with David Brown, a deal which would have given M-H distribution rights for Nuffield or David Brown tractors. Both companies turned down the Massey-Harris offer, and the decision was taken to produce the 744D.

In 1949 production of the 744D was transferred to the newly-opened Massey-Harris factory at Kilmarnock, Scotland. Sales to export markets were satisfactory, and the new tractor helped Massey-Harris to gain a substantial share of the tractor business for the ill-fated groundnut scheme in Tanganyika. On the British market the tractor was a disappointment. During five years of production the total number of 774Ds manufactured was only 17,000. In an effort to increase sales a new version, the 745D, was launched in 1953 with a Perkins L4 engine for greater power and improved fuel economy. The 745D was overtaken in its first few months by the announcement of the merger with the Ferguson organization, and the newly formed joint company switched almost immediately to Ferguson System tractors from the Banner Lane factory in Coventry.

The link between Massey-Harris and the Perkins company of Peterborough, which started with the 744D, resulted in 1959 in the takeover of Perkins by Massey-Ferguson. The first new tractor launched by the group in Britain was the FE35, which was painted grey and carried the name 'Ferguson' above the radiator. The tractor was launched in 1955 with diesel or petrol engines of 37 bhp, or a 30-bhp paraffin engine. The grey colour and the prominent Ferguson identity vanished in 1957 when the paint was changed to M-F red and the tractor became known as the MF35. In the same year a larger model, the MF65, was announced with a 50-bhp Perkins diesel engine.

Work started in 1962 on what was known within the Massey-Ferguson organization as the DX programme. This was a project to develop a new range of tractors to be manufactured and marketed on a completely international scale. Part of the new range was to take the company for the first time above the 100-hp engine range. The programme demanded a million hours of work in the design and prototype evaluation stages. The new range of tractors was launched in 1965, with the larger models allocated to the American factory and the popular smaller and medium-power tractors produced in several locations, including Coventry. The British section of the manufacturing programme included the MF135, 165 and 175 models initially, with derivatives added to the programme later.

A prototype of the 1949 Fraser crawler tractor.

In addition to the transatlantic arrivals in the British tractor industry, there were also plenty of new British companies to compete for a share of the market. Some of these chose to develop crawler tractors, which seem to have appealed rather more to manufacturers than to farmers. Certainly the market failed to develop as the tractor companies had anticipated.

Some of the new crawler tractors appear to have made little commercial progress after the official launching ceremony. Fraser Tractors of Acton, London, demonstrated their prototype model in May 1950. The prototype used a diesel engine 'of continental manufacture', but a British engine was promised in the production version. A year earlier the Glave tractor had been announced by a company at Newport Pagnell, Buckinghamshire. This was equipped with a Morris engine of 12–14 hp.

In its highly specialized sector of the market the Cuthbertson 'Water Buffalo' was a modest success. The prototype was designed for hill and swamp conditions by an agricultural engineering and contracting firm at Biggar, Lanarkshire. Although the prototype was completed in 1949 the first public demonstration was not held until 1951. Then the Water Buffalo was 'driven by a Scotsman paid on piecework

A Loyd tractor on display at the 1948 Royal Show.

rates; and I can think of no sterner taskmaster', reported the journal *Farm Mechanisation*. By 1951 six of the tractors had been built, powered by the Albion EN286 diesel engine developing 70 bhp at 1,800 rpm. A special feature of the tractor was the sealed hull designed to operate in up to 4 ft of water.

The most magnificent of the new arrivals in the crawler tractor market was the VR-180, which was launched in 1952. This tractor combined the engineering resources of the Vickers company of Newcastle upon Tyne with the Rolls-Royce reputation for engines. The power unit was a 12-litre diesel engine developing 180 bhp. The tractor was aimed partly at the farming market, but was mainly intended as a contractor's machine for large-scale land clearance and civil engineering. Production continued until about 1959.

Loyd crawlers were made at Camberley, Surrey by a company which had gained experience of producing tracked vehicles for the army. The first of the agricultural crawler tractors was announced in 1945, and it was followed by a confusing series of new models and engine options until production ended in the early 1950s. The

A narrow-track version of the Howard Platypus with P4 engine.

first version was offered with a V-8 Ford petrol engine, and later with an alternative paraffin version of the same power unit to reduce fuel costs. A third option was the 30-bhp Turner 4V95 diesel, available in 1949. Clutch and brake steering arrived on Loyd tractors for the first time in a new model publicized in 1950 and equipped with a Dorman 4DWD engine. A smaller model announced at the same time was the 105CM, using a Fordson paraffin engine. These new models were apparently superseded in the same year, when the Loyd 'Dragon' was announced at Smithfield Show. The engine options were the Dorman 55-bhp unit or a Turner diesel.

A platypus is a small Australian animal, which is powerful in relation to its size and is at home in wet marshy areas. This was the name chosen by Mr A. C. Howard for the range of crawler tractors produced at Basildon, Essex by his Rotary Hoes group of companies. Mr Howard was an Australian by birth and his 'Platypus' tractors were produced in a range of models which included versions designed for operating in boggy conditions.

The model 28 Platypus was launched at Smithfield Show in 1952 with a choice

Tracklaying conversion of the Fordson Major made in 1949 by County.

of a Perkins P4 diesel engine, or a Standard petrol engine. Later a Perkins R6 of 5·5 litres capacity was offered as an alternative. Platypus tractors were available in standard versions, with narrow tracks for inter-row work, or extra wide track for coping with soft ground conditions. The 'Bogmaster' model was the wet land version, with extra long tracks and a 32-in. track plate to give a claimed ground pressure of 1·3 psi. An ambitious plan to manufacture a load-carrying version of the Bogmaster, to be called the Bogwaggon, never materialized. The design included a forward-control operating position, placing the driver beside the engine, and with the rear of the tractor modified to carry a tipping truck body. This was to be equipped with a spreader unit for distributing soil, lime or other materials.

County Commercial Cars and Roadless Traction both moved into the crawler tractor market at about this time, and both moved out again later to concentrate on the four-wheel drive tractors which have made both companies famous in most parts of the world.

Roadless were the crawler track experts, but most of their experience had been in making tracks for other manufacturers to fit. The 1954 Royal Show was the occasion chosen by Roadless to announce their RT20 tractor. This was equipped with a 20-hp Perkins diesel engine and a transmission giving six forward ratios and two reverse through a transfer box. A tracklaying version of the Fordson Major proved a better commercial proposition, and this was the tractor produced for sale.

County also began production with a full-track version of the old Fordson Major in 1949. A Perkins P6 engine could be fitted instead of the Fordson paraffin engine. A clutch and brake steered crawler version of the new Major was announced in 1951 as the Fordson Major County.

David Brown manufactured some crawler tractors during the war, but their best known model was introduced in 1949 as the 'Trackmaster', based on their successful 'Cropmaster' wheeled tractor. The crawler tractor used the same basic engine as the Cropmaster, bored out to nearly 10 per cent greater capacity. The transmission included a high and low ratio box to give six forward speeds and two reverse. A two-speed power take-off was standard, and this also worked through the twin range box to provide four speeds from 520 to 2,000 rpm. A diesel engine could be fitted from the 1950 Smithfield Show onwards, and twelve months later the Trackmaster 50 series was announced, with a six-cylinder diesel engine as standard.

With so many newcomers arriving to share the distinctly limited crawler tractor demand the manufacturers already established in the market came under considerable competitive pressure. They all announced their own new or improved models during the period after the war, to provide the customer with more choice than he had ever enjoyed previously, or since.

Ransomes announced the MG-5 in 1948, to replace the previous MG-2 model of their small crawler tractor, produced intermittently since 1936. The new version

The MG5 version of the Ransomes crawler tractor.

continued to use the same 600 cc single-cylinder petrol engine as its predecessor. This engine was rated at 4·25 hp, and the drive was taken through a 'self-energizing' clutch, which automatically transmitted the drive as the engine speed brought the flywheel up to 500 rpm. Later versions of the tractor, introduced before production ceased in 1966, were available with an optional diesel engine.

The MG tractor was designed for market-garden work, and a range of special implements was produced to match its small size and modest power. The size of the tractor suggested possibilities which were beyond the scope of a full-size crawler. One MG tractor was used for exploration in Canada on an expedition with a canoe. Depending on the circumstances, the tractor sometimes carried the canoe and the explorer, or alternatively the man and the tractor could travel in the canoe. There was some interest, also from Canada, in equipping MG tractors with fire-fighting equipment, so that they could be dropped by parachute to deal with forest fires in remote areas.

Although the fire engine project was abandoned, Ransomes found many other export markets for the tractors. In Tanzania they were used to harvest salt. Wooden

Bristol 20 crawler photographed for the NIAE reference collection, 1949.

extensions helped to spread the track pressure, and the tractors were used both to scrape up the salt from the surface of the salt-pans, and also to remove the salt using a dumper conversion. In Holland some farmers used MG tractors for field work, transporting the tractors across drainage canals on boats which were modified so that the propeller was powered from the tractor power take-off (p-t-o).

Bristol tractors also appeared first in the 1930s, intended originally for the market gardener. After World War II the Bristol organization was taken over by H. A. Saunders, the Austin car distributors, and Austin engines were fitted. The principal postwar model was the 20, using a 22-bhp Austin industrial engine, based on the power unit for the Austin 16 car, but with a paraffin conversion. In spite of their name, Bristol tractors were produced after the war in a factory at Colne, Lancashire. In 1953 the popular Perkins P3 diesel engine was offered as an option.

Before the Platypus tractor had been produced by Howard, John Fowler of Leeds had been temporarily a subsidiary of Mr Howard's Rotary Hoes group. For this reason the first new Fowler crawler tractor produced after the war, the FD2, was launched with a special side-drive Howard Rotavator, operated from a p-t-o point at the side of the tractor.

77

The Fowler FD2 was powered by a four-cylinder Fowler diesel engine, giving 24 bhp, and driving through a range of six forward and six reverse gear ratios. The new tractor arrived on the market in 1946, and survived only two years in production. The Fowler company merged with Marshall of Gainsborough as part of the T. W. Ward group, and the first production tractor to result from the merger was the Fowler Mark VF, which linked a single-cylinder Marshall diesel engine, and Marshall styling, with Fowler track know-how. The slow-revving two-stroke power unit, with its 6·5-in. cylinder bore, was already somewhat outdated, but was well known and respected.

An improved version of the VF arrived in 1953, distinguished as the VFA, but still retaining the same engine design. Meanwhile additional crawler models had been introduced in the Challenger series which went into production in 1950. These were heavier and more powerful than the VF type, and used more up-to-date engines, including the 95-bhp Meadows diesel in the Challenger Mark III at the top

Fowler Mark III Challenger, 1951.

78

The Bean toolbar, designed for precision work in rowcrops.

of the range. The Mark 11 and Mark 111 Challengers arrived on the market in 1950, and were followed twelve months later by the Mark 1 which was powered by a 50-bhp twin-cylinder two-stroke diesel engine made by Marshall.

Some of the more innovative designs appeared in tractors designed for inter-row work on market gardens and other high-value crop production. An early arrival in this category was the Bean toolframe, designed by a large-scale vegetable producer to suit his own requirements. It was manufactured in Brough, Yorkshire, from 1946, and later by Green of Leeds. The Bean was designed around a rectangular framework mounted on three wheels. The two wheels at the rear were driven by a Ford 8-hp industrial engine, and the single wheel at the front was tiller-steered. The engine was mounted at the rear directly over the wheels and out of the line of vision of the driver. The driver's seat was in the middle of the framework, and the tools, such as spray units, seeders, hoes and scufflers, were located directly in front of the operator so that he could control them precisely.

The Bean tractor was a good example of simple but effective design, and it was awarded a silver medal by the RASE at the 1947 Royal Show. Competitors included the Wild Midget Tool Chassis, which was also based on a simple metal framework mounted on three wheels. The general layout of rear engine, mid-operating position and forward position of the toolbar mounting was similar to that of

Newman three-wheeler of 1948.

the Bean. The Wild version had a single driving wheel at the rear and two steering wheels at the front. The power unit was a cheaper JAP 4·5-bhp petrol engine, and the driver's seat was an economical strip of canvas.

Other row-crop tractors or self-propelled toolbars included several designs which were obviously forerunners of the David Brown 2-D. The Opperman Tractivator, launched at the 1948 Royal Show with a Douglas 8-bhp engine, was of this type. The Gunsmith was another example, also launched in 1948, and the very much more successful Garner tractor announced in 1949.

New arrivals on the tractor market in 1949 included a range of small tractors from Newman Industries of Grantham, Lincolnshire. These were designed to take an underslung, mid-mounted toolbar for rowcrop work, as well as working with trailed equipment. Engines for the Newman range included a 567-cc Coventry Victor diesel, and 10·75-bhp and 12-bhp air-cooled petrol engines by the same maker. Selling prices for the Newman tractors ranged between £240 and £330 for the basic tractors in 1949, to sell against a basic Fordson listed at £237 upwards.

Similar problems of price faced the Byron tractor, selling in 1949 for £247, or £292 in rowcrop form. The Byron was made in Walthamstow, London, and like the

Byron tractor fitted with underslung toolbar.

Newman was designed for both drawbar and rowcrop work, fitted with a Ford 10 engine.

At about the same time, and with the same type of engine, the OTA tractor was introduced by a company called Oak Tree Appliances of Coventry. This tractor started as a tricycle design, with a single front wheel, but a more conventional four-wheel layout was adopted from 1952. A three-point linkage was offered, and the basic three forward ratios in the gearbox were doubled by the later addition of a transfer box.

Other tricycle designs included the Kendall tractor and the Ford-engined Power-steer. The latter, announced in 1949 by Maxim Engineering of Ladbroke Grove, London, was unusual in using the rear wheels for both driving and steering. The front wheel, impracticably tiny on the prototype, was simply a castor wheel. Much more successful was the President tractor launched at the 1950 Royal Show the resources of the Brockhouse group. This was a four-wheel design, with p-t-o and three-point linkage options, and with a Morris 8/10 engine in petrol or tvo (tractor vaporizing oil) versions.

Fortunately the development and production of tracklayers and small tractors was not the only activity in the industry during the postwar period. There were also

81

The OTA tractor, 1951 version.

BMB President tractor fitted with Howard Rotavator.

Turner 'Yeoman of England' photographed in 1949.

important new makes and models in what were then the medium and high horse-power sectors of the wheeled tractor market. These included several new four-wheel drive models, a type previously neglected by British manufacturers.

One of the new entries at the upper end of the market was the Turner Manufacturing Company of Wolverhampton. Their first tractor was launched at the 1949 Royal Show, and attracted considerable attention because of its unusual engine design and comprehensive specification. The company later named the tractor the 'Yeoman of England'.

The power unit was produced by Turner, and was a diesel engine with four cylinders in a 68° V-formation. The engine was rated at 34 bhp at 1,500 rpm, with a maximum output of 40 bhp. The tractor was available with hydraulic lift, p-t-o, three-point linkage, electric lights, electric start, adjustable track width, and with the steering wheel and driving position slightly offset from the centre line for better forward visibility. A differential lock was an optional extra. A slightly modified version of the tractor appeared in 1951, with detailed engine improvements, and the Yeoman remained in production until 1957.

Design work for the new Nuffield tractor began as soon as the war had ended. Lord Nuffield controlled one of the giant companies in the British motor industry, and the new tractor was planned to fill some of the manufacturing capacity which

Nuffield 4-60 tractor with Allman sprayer attached.

peace and the end of military contracts would leave vacant. By 1946 the first proto-type tractor was working on a Lincolnshire farm, and during the same year a further eleven prototypes were built for field testing.

The new tractor made its first public appearance at the 1948 Smithfield Show. The standard version of the tractor was the M4, with a tricycle version, known as the M3, available for rowcrop work. The tractors were manufactured at first in the Wolseley car factory in Birmingham, using a Morris Commercial ETA engine developed for army vehicles. The engine was a four-cylinder petrol/paraffin unit, with a maximum output of 42 bhp at 2,000 rpm, and 35 bhp at 1,400. The first diesel version was available from 1950, using a Perkins P4, which was replaced in 1954 by a 3,402-cc diesel engine manufactured by the British Motor Corporation, of which the Nuffield companies were then a part.

A smaller Nuffield tractor, the Universal 3, was added to the range in 1957, when the larger model became known as the Universal 4. Both tractors were improved in 1961, after production had been transferred to the Morris factory at Cowley, Oxford. The new versions were first shown at a press demonstration near Oxford in November, in preparation for a Smithfield Show launch. The Universal 4 became the 4/60, and was described as 'Britain's most powerful tractor'. The claim was based on an increased engine capacity, from 3·4 litres to 3·8, with the output raised to 60 bhp at 2,000 rpm.

Another face-lift in 1964 brought in the 10/60 tractor, with the same engine as its predecessor but with a double-ratio gearbox to give ten forward gears and two reverse. The smaller tractor in the range was similarly improved, including disc brakes and better hydraulics, to become the 10/42. Nuffield tractor production had already been transferred in 1963 to the new factory at Bathgate, Scotland. There the Universal series ended its production life in 1967 when the new-look Nuffield tractors were announced at the Royal Show. Four years later the Nuffield name disappeared from the market when a new range of tractors, carrying the Leyland

A David Brown VAK 1 at a wartime demonstration at Fladbury, Worcestershire.

name and with a new blue colour scheme, arrived from a company which had become part of the British Leyland group.

The first tractor designed, built and marketed by David Brown was the VAK 1, which arrived in 1939, shortly before the outbreak of war. The new tractor included several important changes which the David Brown company had wanted to introduce on the tractor they had previously made for Harry Ferguson. Ferguson had been unwilling to accept the need for changes to his original design, and it was the VAK 1, produced after the break between Ferguson and the David Brown company, which included the four-speed gearbox and the bigger engine which David Brown considered necessary. The new design was also equipped with a proper power take-off shaft and was given more modern styling. Patent restrictions prevented the David Brown designers from using the full Ferguson System, and a depth-limiting wheel was used as a substitute for automatic depth control.

After the war years, when David Brown factories had been producing tractors for towing work with the Royal Air Force and Royal Navy, an updated version of the VAK 1 tractor was produced as a temporary measure while a new model was being developed. The VAK 1A, as the first postwar tractor was called, included a number of detail refinements, such as better lubrication and an improved engine governor. In 1947, after a total of nearly 9,000 VAK 1 and 1A tractors had been produced, the new and highly successful 'Cropmaster' series was launched.

The Cropmaster included an optional six forward ratios and two reverse, engine improvements and an improved three-point linkage with lift capacity increased to 1,330 lb. at the ball ends. A diesel version, using the same basic engine, was announced in 1949, and in the following year a more powerful export version, known as the 'Super Cropmaster', was on the company's Smithfield Show stand. This was again updated in the following year as the Cropmaster diesel 50, again described as being mainly for export. This version of the basic trackmaster was fitted with the well-proven Perkins P6 engine.

Production of the various Cropmaster series tractors ended in 1953 after almost 60,000 had been manufactured, and David Brown had become firmly established as one of the major international tractor companies.

The 950 model, introduced in 1958, and the smaller 850, which first appeared in 1960, were both manufactured between 1960 and 1963 under a special arrangement with the American Oliver Corporation. David Brown produced the tractors in Oliver green, and shipped them to the United States for sale there as part of the Oliver range. Meanwhile David Brown were still using red paint, or 'hunting pink' as they preferred to call it, for their own tractors, and this was changed in 1965 to the now familiar white, which makes David Brown tractors among the most distinctive in appearance.

In 1972 David Brown Tractors Ltd was bought by the American Tenneco com-

Field Marshall Series 3 in 1951.

pany of Houston, Texas. Tenneco already owned the J.I. Case company, and the take-over of David Brown gave the British-built tractors additional marketing strength in the North American market.

One of the first of the new tractor models to be announced after the end of World War II was the Fordson Major, the famous E27N tractor which replaced the ageing Model N Fordsons. The Major was coming off the production line in the summer of 1945, available in both standard and rowcrop versions. The retail price for the standard version on metal wheels was £237, or £285 on rubber tyres. The

Marshall MP6 tractor with Leyland engine, 1956.

Fordson Major with tvo engine, 1952.

County Four Drive tractor with skid steering.

Doe Triple-D articulated tractor of 1960.

rowcrop model on metal wheels was listed at £255. The engine was rated at 28·5 bhp at 1,100 rpm on paraffin, with 19·1 hp available at the p-t-o.

The Fordson Major, with later diesel versions, held the Ford position in the market while a completely new model was being prepared for the production line. This was the new Major, launched at the 1951 Smithfield Show to become one of the most important and successful of British tractors. The new Major offered a particularly wide range of options. There was the choice of diesel, petrol or tvo engines, and wheels, half-tracks or, in the County version, full crawler tracks.

The engines were four-cylinder ohv (overhead valve) designs, rated at 40 bhp, at 1,700 rpm. The diesel version was 3,610-cc capacity, with a 16 : 1 compression ratio. The tvo engine was the same capacity as the diesel, but the compression ratio was only 4·4 : 1. The petrol engine had a 5·5 : 1 compression ratio and was 3,261 cc. Three-point linkage and power take-off were optional extras, but the double-ratio gearbox with six forward speeds was standard.

It was the diesel engine in the new Fordson which made the greatest impact. It was an easy starting engine, and it helped to encourage the trend away from petrol and tvo. This was a trend which developed quickly in Britain, helped also by the success of the Perkins P6 and P4 engines, and tractor manufacturers in Europe and America were forced to offer diesel engines in order to compete.

Ford improved and developed the original new Major tractor during its production life, offering extra power in the Power Major and Super versions. Less successful was their entry into the lower-horsepower end of the market with the 'Dexta', which was always overshadowed by the much more popular Major series.

All the large-scale tractor manufacturers in Britain offer four-wheel-drive versions, or four-wheel assist, of their more powerful models. This has been one of the most significant design trends during the 1970s, to some extent influenced by the increasing sales of the four-wheel-drive tractors already available in Europe. Previously in Britain four-wheel drive had been regarded as a specialized sector of the tractor business which the big companies considered too small to be of interest.

This left a gap in the market which was exploited with considerable success by companies such as Bray, County, Doe, Muir Hill and Roadless. Using engines or skid units from the larger manufacturers, the four-wheel-drive specialists developed and produced their own tractors, offering extra traction efficiency in exchange for a more expensive price. Bray specialized in Nuffield conversions, Muir Hill used engines from Ford and Perkins, and County, Doe and Roadless built up their businesses on Ford skid units.

Roadless was already well known as a crawler specialist, making half-track conversions and a crawler conversion of the Fordson Major before switching to four-wheel drive. The Roadless designs are similar to most of the European four-wheel drives in having smaller wheels at the front. This is sometimes called four-wheel assist, and simplified some design problems, including steering and price.

The County preference is for four equal-size wheels, which is claimed to give more efficient traction. This was the design chosen by County when they produced their 'Four Drive' tractor, intended for use in the West Indies for harvesting sugar-cane. The Four Drive went into production in 1954, with an unusual skid system of steering which was developed from the County crawler tractors. Although four-wheel drive proved to be commercially successful, the skid steering arrangement was too limited, and more conventional steering was chosen for subsequent County tractors.

Another of the more adventurous designs from County was the 'Forward Control' tractor, produced in 1965, and more recently updated as the FC1174, which was launched in 1977. The Forward Control offers the advantages of four-wheel drive in a tractor of 100 hp plus, with a cab position designed for outstanding visibility, and a space behind the cab which can be used as a load-carrying platform.

Ernest Doe and Sons of Maldon, Essex, is one of Britain's largest tractor and farm machinery distributors. They moved into tractor production in 1957 to meet a demand for extra power and pulling efficiency on the heavy land arable farms in their sales area. Their new tractor, the Triple-D, was based on the design of an Essex farmer, and consisted essentially of two Fordson Major tractors joined to-gether. The front wheels and axles were removed from both tractor units, and the front end of the rear tractor was hitched to the rear of the front tractor. The result was a twin-engined tractor unit on four large driving wheels, and with both engines operated by a single set of controls.

The Triple-D offered British farmers the manoeuvrability of an articulated trac-tor for the first time, and also an opportunity to buy nearly 100 bhp. The Doe tractors had some disadvantages, including their excessive length and doubling the number of most components. But they filled a real gap in the market, selling as ploughing tractors in Britain, and for heavy transport work in West Africa and Scandinavia. More than three hundred of the tractors were built, including a small number of the later 130 model, based on the Ford 5000 and offering nearly 130 bhp.

Special-purpose Tractors

Much of the effort which has so far gone into the development of the tractor has been aimed at providing more versatility. The first tractors, direct descendants of the steam traction engine, were used mainly for stationary work. Horses were the main source of power, and it was not until the turn of the century that the tractor was sufficiently versatile to become a replacement for the horse.

Since that time engineers in many countries have contributed the improvements which have transformed the tractor into its modern versatile form, suitable as the power source for a varied range of operations. This is the tractor which most of the world's mechanized farms require, with the Jack-of-all-trades adaptability to cope with all the varied jobs throughout the year.

However, there is also a need for tractors designed to do only a limited range of jobs, but to do them with extra efficiency. Sometimes these are simply modifications, retaining much of the versatility of the original general-purpose design. Examples include four-wheel-drive versions, designed to give extra efficiency at the drawbar, and the rowcrop tractors popular in some arable farming areas.

Specialized tractors also include some which break away completely from versatility and conventional design, in order to offer high output or greater efficiency in their specific roles. Recently this type of specialization has been attracting increased interest. An example is the rough-terrain forklift or materials-handling tractor, adapted for farming from its industrial beginnings and now an important factor in agricultural mechanization.

British farmers welcomed the forklift, and demand has provided a market large enough to encourage a thriving home-based production, plus imports, particularly from France. Arable farmers were the first to buy rough-terrain forklifts, but there is also a demand from the larger grassland farms, providing a total market amounting to more than one thousand units a year in the late 1970s, and still increasing in Britain.

The familiar type of forklift is a two-wheel-drive unit, with the engine at the rear, small wheels also at the rear for steering, and with the driver positioned towards the front of the unit. The large driving wheels are also at the front where the weight is carried during work. The forklift mechanism consists of a mast, and a carriage which moves up and down the mast. The carriage is moved vertically by hydraulic pressure, and is the attachment point for the various tools which may be used for materials-handling operations. Attachments available include scoops for dealing with slurry or loose granular materials, forks for silage, grass and manure, root-handling buckets, bale clamps and flat-8 bale grabs, box rotators and tipplers, and forks for pallet handling.

To improve the performance of the basic forklift unit there are numerous developments and options. These include four-wheel drive, four-wheel steering, power steering, the ability to tilt the mast for more precise operation, and to move the carriage and its attachments sideways on the mast as well as vertically. Torque converter transmissions are available instead of a manual gear change, and there are special high-reach masts, and masts for working inside buildings with low roof clearance.

The rough terrain forklift is designed almost entirely for materials-handling work, and is a good example of a highly developed special-purpose tractor. A conventional farm tractor fitted with a front-end loader or with a forklift attachment on

Bonser four-wheel-drive materials-handling tractor working on a silage clamp.

its three-point linkage will do all the jobs which a rough-terrain forklift can deal with, but with less efficiency. British farmers have led the world in accepting the advantages of the materials-handling tractor, and in benefiting from its greater efficiency for lifting, loading, stacking and moving. The rough-terrain unit keeps its load close to the front axle for greater lift capacity and stability, and also close to the driver for more accurate control. The rear steering and side-shift on the carriage mean that loads can be placed with precision. Compared with a tractor and front loader the farm forklift is more compact for manoeuvring in confined spaces.

Among British manufacturers, Bonser, Matbro and Sanderson were all quick to establish themselves in the agricultural market for forklifts, competing against a growing list of manufacturers, including Manitou, Salev and Sambron from France. A more recent arrival in the market is JCB, with a different type of materials-handling tractor. Instead of a forklift, JCB offer an extending boom on their tractor. This was launched first in the construction-equipment market, where JCB is one of the best-known names. It was quickly realized that the same idea might

Sanderson forklift with bale carrier, 1978.

94

have a place on the farm, and the JCB 520 arrived in the agricultural market in 1978.

The basic layout of the JCB unit is similar to that of a forklift, with rear-wheel steering, front-wheel drive, and with the driver located between the axles in a position with good all-round visibility. The materials-handling equipment, such as bale clamps and manure or silage forks, attach to the end of the boom. The boom is hinged from its attachment point over the rear axle and is raised or lowered, like the jib of a crane, by two hydraulic rams. The rams lift from a point over the front axle, so that the tractor has good stability in work.

A distinctive feature of the JCB materials-handling tractor is that the jib or boom is made up of a number of telescopic sections. These push out under hydraulic pressure to extend the boom and its load outwards and upwards. In some situations this can give greater versatility than the rough-terrain forklift can achieve. It means that a load of bales can be raised to the top of a stack, and then reached across to the far side of the stack. Loads can be spread more efficiently, even on a high-sided trailer, and the boom can be extended over the sides of a pen or yard which the tractor unit does not have to enter. This appears to add, literally, a new dimension to materials handling on the farm.

Demand for crawler tractors increased in Britain during the 1950s, but the modest sales boom was short-lived. One of the main reasons for the renewed interest in the crawler was the extra drawbar pull it offered, and wheeled tractors became increasingly competitive as they offered more powerful engines and four-wheel drive.

The rival merits of the tracklayer and the wheeled tractor have been argued for years, and the debate will probably continue. Some of the rivalry concerns operating

JCB 520 materials-handling tractor, 1978.

costs. This is difficult to resolve because costs for tyres and tracks depend upon driving habits and soil type and condition, and these are all variable. Work outputs are also difficult to compare fairly, as the faster working speed which may be possible with a wheeled tractor must be measured against the possibly greater pulling power of the crawler, and ground conditions again influence the answers.

Wheeled tractors inevitably offer greater versatility, and the fact that tracks must be kept off public roads is a serious disadvantage on some farms. The crawler also has some considerable advantages, and these may become increasingly important in the future. One of these advantages is greater pulling efficiency than a wheeled tractor of similar engine power, and this also applies when the wheeled tractor is four-wheel drive. The gain is likely to become most significant on heavy soils, and especially in wet conditions. As long as fuel is relatively cheap it can be preferable to increase drawbar pull simply by buying more horsepower. Rising fuel costs will put more emphasis on using the power efficiently.

Another major gain for the crawler is that tracks are less likely than wheels to cause damage to soil. One of the reasons for this is that the weight of a crawler tractor is spread over a large area of ground. The wheeled tractor is no match for the 6·5 lb. per square inch – or less – ground pressure of a medium-sized crawler. This is the reason why many arable farmers prefer to use a crawler tractor for jobs such as seedbed preparation and seed drilling, to minimize the risk of compacting the soil and inhibiting crop development.

Tractor tyres can also damage soil if wheels are allowed to spin during ploughing or other drawbar work on wet land. Wheel-spin produces a smeared surface with compaction below, and this can cause reduced crop yields, interfere with natural drainage and hinder future cultivations. A tracklayer is able to operate in a wider range of conditions without causing smear, and this can be a significant gain on a heavy land farm. As the crawler tractor may be able to operate in conditions where a wheeled tractor would cause damage, the crawler can often put in more working hours at critical times of the year and save time significantly.

Britain's stake in the crawler market has been held by Aveling Marshall, part of the Leyland organization, and variously known in the past as Track-Marshall and Marshall-Fowler. This company stayed in the crawler business while most of its rivals turned to more glamorous markets. It was Marshall-Fowler which brought the benefits of a proper hydraulic three-point linkage to the crawler, and helped to introduce the power take-off.

Leyland sold their interest in the agricultural crawler tractor business in 1979, and Aveling Marshall tractors are now owned by a Lincolnshire farmer who is continuing production in part of the old Marshall factory at Gainsborough.

One of the specialist jobs for which the conventional tractor is not ideal is inter-row cultivation. Specially designed tool carriers allow much greater precision for

Aveling Marshall 100 crawler.

working between rows of plants, to help control weeds without damage to the crop.

Tool carriers must be designed to give the driver precise control over steering and a good range of slow working speeds. The driver must have the clearest possible view of his work, and the toolbar to which implements are attached is usually mid-mounted so that the driver can watch them constantly with little effort. Light weight, to avoid soil compaction, and good manoeuvrability to minimize turning space, are also important design features.

The most successful British tool carrier came from David Brown when they launched their 2D model at the 1955 Smithfield Show. This was powered by an air-cooled, twin-cylinder diesel engine of 14 hp, driving through a four-speed gearbox. The 2D was designed to work with a mid-mounted toolbar, but front and rear hitch points were also provided. An ingenious design feature was the use of compressed air for raising and lowering the toolbar, with the tubular metal frame of the tractor acting as the air reservoir.

David Brown 2D tool carrier, 1958.

The 2D went out of production in 1961, after 2,008 had been built. Some were still in regular use twenty years or more after they were made. In 1978 several companies realized there was a market for a new tool carrier to replace the ageing 2Ds. One of the British companies to move into the market was Russell of Kirbymoorside, York. Their new tool carrier was designed specifically as a replacement for the 2D, and was called the 3D, by arrangement with David Brown.

A two-cylinder Lombardini engine is used on the Russell tractor, driving through a hydrostatic transmission to give infinitely variable speeds up to 9 mph in both forward and reverse. The toolbar is controlled hydraulically, and accepts many of the original 2D attachments.

Trantor is an attempt to break away from conventional ideas and to provide a tractor with special advantages for transport work. The project started in 1971 after Stuart Taylor, a post-graduate university student at the time, carried out a research investigation into the use of farm tractors in Britain. The data he collected showed that many tractors spend more than 50 per cent of their working hours on transport and travelling. The conventional tractor, Mr Taylor decided, was designed to give

Russell 3D tool carrier with hydrostatic transmission, 1978.

its best performance at ploughing speeds and was far from ideal as a transport unit.

He formed a small company, W. S. H. Taylor Engineering Developments, at Heaton Mersey, Stockport, to develop his ideas for a new tractor. His objective was to build a tractor which could still be used for general work with a wide range of implements, but which offered a high performance for travel and transport, both around the farm and on the road.

The first of Stuart Taylor's tractors, under his Trantor trade name, arrived on the market in 1978. The specification included a top speed on the road of about 65 mph, with steering and brakes capable of coping safely with this performance.

One job which Trantor does better than a conventional tractor is carrying people. The cab is an integral structure with the welded-steel box section chassis, and the leaf springs which form the rear suspension, give an acceptably comfortable ride across rough ground at speeds of 25 mph or more. The cab has a central driving

99

seat plus seats for two passengers. Seats for four more passengers can be fitted at the rear of the cab, instead of the alternative load-carrying platform.

For trailer work on the road Trantor has a hydraulically operated hitch unit with a swinging drawbar. To cope with the shock loadings which may occur at speed, the drawbar is protected by a massive transverse leaf spring. This can accept trailers with a load capacity up to six tons, and with 40 per cent drawbar loading, while allowing a comfortable suspension for the cab and chassis. A three-point hitch is included in the drawbar unit, and there is a power take-off as standard equipment. The hook on the end of the drawbar is positively latched when in use and is linked to a safety warning light on the dashboard.

Many of Trantor's design features are closer to the standards of a truck than a tractor. The steering is of the recirculatory ball type, and there is air-assisted braking to all four wheels plus the facility to take air braking to the wheels of a trailer. The front suspension can be locked rigid when working with a front-end loader, and the rear wheels can be braked individually to assist manoeuvrability.

Early versions of Trantor were fitted with either a Leyland engine developing 75 bhp or a Perkins unit of 78 bhp. The gearbox is based on a Bedford truck unit, but with a transfer box to give eight forward ratios and two reverse.

One of the theories behind the Trantor project was that the tractor should be designed for manufacture or assembly – in quite small numbers if necessary – overseas, including developing countries. Stuart Taylor's group backed Trantor with a service to advise on the design of Trantor production factories, including training the workforce and providing the equipment.

The first response to Trantor has been hesitant. Some of its critics have concentrated on its lack of suitability for ploughing, rather than on its advantages for the work it was specifically designed to do. Stuart Taylor emphasizes that his design was not planned to compete with the heavy ploughing tractors on the farm. He also points out that materials-handling tractors have been welcomed although they offer considerably less versatility than Trantor.

British engineers have achieved considerable success in helping to meet the requirements of agriculture in widely varying conditions abroad, and the British tractor industry now relies heavily on export business. During the 1970s considerable efforts have been made to design a simple, inexpensive tractor for developing countries, a project which could one day improve the lives of millions of the world's poorest people.

The problems to be overcome are immense. The countries for which this type of tractor is being developed lack the money to mechanize with existing tractors and equipment, which means that a specially designed mechanization system must be cheap. The equipment must be suitable for use in areas where servicing and operating skills are minimal. It must also stand up to working in difficult conditions,

An experimental tractor from the NIAE in 1965, with hydrostatic transmission and a driverless guidance system.

Buffalo tractor designed for developing countries, 1977.

including dust and cultivating hard-baked soils. Ideally it should be designed for export in kit form for assembling locally on a small scale by relatively unskilled labour, as this would help to reduce the cost and would also provide some employment in the importing countries.

Much of the development work has been carried out at the National College of Agricultural Engineering at Silsoe, Bedfordshire, and by the NIAE, also at Silsoe, with financial assistance from the Ministry of Overseas Development. Among the various systems which have been evaluated are several which use a winch to pull implements through the soil. A small tractor unit is used which carries a winch and cable. At the end of the cable is a plough. The tractor unit propels itself forward the length of the cable, and is then anchored while the winch is engaged to pull the implement up to the tractor. This system makes efficient use of a small engine, especially in difficult soil conditions. The fact that the work rate is slow is not a major problem in countries where the aim is not to displace large numbers of people from the land.

One of the prototypes built by the National College is code-named 'Spider'. It is a simple, fabricated chassis with a single, tiller-steered front wheel and two powered rear wheels. The power unit in the prototype is a 6·5-bhp diesel, driving through a vee-belt and chain reductions. The winch is mounted on the rear axle of the tractor and is used for the more demanding draught jobs. The Spider has adequate pulling capacity for light transport work with a trailer and for secondary cultivations. The tractor unit could also be used as a stationary power source for work such as pumping water or generating electricity.

During 1977 a prototype third-world tractor was produced for commercial evaluation by Industrial Engine Sales of Grantham, Lincolnshire, a company within the Elbar Industrial Group. Much of the design work was the responsibility of Barford of Belton, Lincolnshire, and it is likely that Barford would manufacture components if the project proves to be a commercial proposition.

This Lincolnshire tractor is known as the 'Buffalo', and is a four-wheel design powered by a Lister 10·5-bhp air-cooled diesel engine, with a three-speed gearbox. The tractor is designed for drawbar work as a replacement for oxen, and could be exported either fully assembled or as a kit. At the time of writing commercial production had not started.

Tractors for World Farming

Few industries can match the trading record of Britain's tractor manufacturers. Export business has been developed with outstanding success, earning large amounts of foreign currency to help the balance of payments. At the same time export sales have made a substantial contribution to the development of mechanized farming in many parts of the world.

The measure of the success achieved by manufacturers in Britain is highlighted by the trade statistics. In the thirteen years from 1966 to 1978 Britain exported almost 1·5 million tractors, including both wheeled and tracklaying models. Imports during the same period amounted to less than 100,000. Over the thirteen years the level of export business often reached as much as 80 per cent of all the tractors

David Brown 1210 tractor at a Bigbale handling demonstration in 1978.

produced in Britain. In addition there has also been an important export trade in used tractors, amounting to 170,000 during the period.

Britain exports large quantities of tractor parts, which are used for tractor production in overseas factories as well as for the existing population of tractors already on farms. Engines for tractors, and for combine harvesters and other farm machinery, are also an important export item, worth £80 million in 1978, compared to £23 million for tractor engines imported in the same year.

One further form of export business is the sale of the equipment and know-how to set up tractor production abroad. A recent example was the £150 million deal through which a consortium of British companies, headed by Massey-Ferguson, provided Poland with a new tractor-making industry for up to 75,000 tractors a year.

Some of the overseas customers for British tractors are developing countries where farming may be still on the fringes of mechanization, and where each new tractor may replace hours of toil. At the other extreme are countries where farming

Massey-Ferguson 265 tractor, 1979.

International Harvester 484.

is already highly mechanized, and where British exporters may be competing with established domestic tractor manufacturers.

Britain's position in world tractor markets is strong and established, but there is growing evidence that competition is increasing, and British companies will find difficulty in future in maintaining recent export volumes. Some of the competition comes from Russia and Eastern Europe, where tractor production has been expanded with the export trade in mind, and where American and British licensing agreements have often helped to put the manufacturing industry on to a modern footing. Tractor sales from Eastern Europe are sometimes linked to aid and development programmes, or may be intended simply to earn much-needed foreign currency. In either case the prices charged probably have no direct relationship to production costs.

Japanese and EEC manufacturers have also become increasingly competitive. The Japanese established their industry on the small tractors which were ideal for their domestic market of small farms, and which also filled the gap at the bottom end of the tractor market obligingly left vacant by most American and European manufacturers. From this highly successful base the Japanese manufacturers are now moving up the market to compete more directly with American and European products. There is also a market sector at the top end of the power range for tractors of 200 hp or more, where it has not been economical for British manufacturers to compete, which has been dominated by American and Canadian companies.

Additional problems in future export trade include uncertainties over fuel costs and the security of supplies, lack of political and economic stability in some markets, and the policy in some countries of sacrificing much-needed spending on agricultural development for the sake of military investment. There is still a huge potential market for tractors, which can develop only when there are suitable economic and political conditions.

While competition increases in export markets there is little room for complacency among manufacturers in their home market. Numbers of tractors sold each year in Britain have been tending to fall, although the average horsepower has been

Leyland 472 four-wheel drive with change-on-the-move transmission.

increasing. At the same time imported tractors have been taking an increasing share of the British market. Part of the increase in imports is a result of the policy of the big multi-national companies to produce only a section of their total range at each factory. This means that even the biggest exporters from Britain must also import some tractors in order to offer British farmers an adequate choice. Imports have also increased because of the success some overseas manufacturers have achieved in establishing themselves on the British market. Some of the imported makes are backed by well-established parts and service availability and strong dealer networks, and these manufacturers expect to increase their share of the market.

Although market conditions throughout the world are likely to become more competitive, British-made tractors are well placed to continue their success story in export trade. One of the strengths of the industry in Britain is its structure, consisting of a mixture of the multi-national giants co-existing with a large number of smaller companies specializing in four-wheel-drive, materials-handling tractors,

Ford 6700, 1977.

Muir Hill 121 with Bomford chisel plough.

tracklayers and tool carriers. This gives an exceptionally comprehensive coverage to meet the varied requirements of farming throughout the world.

British exporters also have an advantage in being firmly established in every potential market for tractors, with a reputation based on years of success, and with experience of the widest range of operating conditions.

Above all, manufacturers in Britain have the resources to provide leadership in technical development and design. There is also increased collaboration between manufacturers in Britain and the National Institute of Agricultural Engineering at Silsoe, now with a more commercial policy of operation.

Much of the progress in power farming throughout the world has been a result of British inventiveness. British steam-engine manufacturers a century ago were building their export business on a reputation for leadership in design and development. At the turn of the century men like Albone and Scott led the world in the early development of the tractor. Much more recently the materials-handling tractor and the high-speed transport tractor have shown that the ability to pioneer new ideas to improve the efficiency of farm mechanization is still alive in Britain. There

County 1174 forward-control tractor with rear load-carrying space.

is also progress in the development of a cheap, third-world tractor.

Agriculture faces immense challenges in the future, with economic and political uncertainties and the possibility of an energy crisis, in a world where millions are without sufficient food to eat. The farming industry of the future will demand more new ideas, and the British tractor industry's prosperity will depend upon providing them.

Index

**Figures in bold type refer to pages on
which the items are illustrated**